SpringerBriefs in History of Science and Technology

More information about this series at http://www.springer.com/series/10085

Ido Yavetz

Bodies and Media

On the Motion of Inanimate Objects
in Aristotle's *Physics* and *On the Heavens*

Springer

Ido Yavetz
Cohn Institute for History of Science
  and Ideas
Tel Aviv University
Tel Aviv
Israel

ISSN 2211-4564   ISSN 2211-4572   (electronic)
SpringerBriefs in History of Science and Technology
ISBN 978-3-319-21262-3   ISBN 978-3-319-21263-0   (eBook)
DOI 10.1007/978-3-319-21263-0

Library of Congress Control Number: 2015943847

Springer Cham Heidelberg New York Dordrecht London

Printed on acid-free paper

Springer International Publishing AG Switzerland is part of Springer Science+Business Media
(www.springer.com)

For Tomer, Amalia, and Dafna

Foreword

The purpose of this essay is to present a theory of the motion of inanimate bodies based exclusively on principles laid down by Aristotle in his *Physics* and *On the Heavens*. As the introduction will explain in greater detail, there is no pretense here to capture the definitive reconstruction of Aristotle's theory of bodies in motion—an impossible task in my opinion, given the information we currently possess. It is, however, a plausible reconstruction, albeit not the only one possible. The essay is meant primarily for historians of science interested in the evolution of physics from antiquity to the scientific revolution of the seventeenth century, and to students of Aristotle's natural philosophy. In order to keep a sharp focus on Aristotle's texts, and to remain within the useful format of the SpringerBriefs Series, discussions and references that pertain to the vast literature on Aristotle's natural philosophy have been cut down to a bare minimum.

Initial motivation for this project originated in a graduate seminar I gave jointly with Prof. Rivka Feldhay on Galileo's *De Motu* at the Cohn Institute for History and Philosophy of Science and Ideas in Tel Aviv University. In his *De Motu*, Galileo attempted to formulate principles of locomotion irrespective of more general ideas regarding natural change. Aristotle's *Physics* discusses nature as the realm of all material changes, of which locomotion is but one form. *On the Heavens* discusses aspects of locomotion to the extent that they pertain to general cosmology. Aristotle's surviving works do not contain a separate treatise dedicated to locomotion alone. By contrast, Galileo's study of terrestrial motion, from the *De Motu* to its final evolved form in the *Discourses on Two New Sciences*, reveals no interest in the wider scope of Aristotle's terrestrial natural philosophy. Instead, he developed a theory of locomotion for its own sake. However, in the course of formulating his own ideas, he found numerous occasions to comment at length on various aspects of Aristotle's theory of locomotion. On some of these occasions, Galileo's comments seemed to us either misguided, or somewhat careless, or not representative of a decent interpretation of Aristotle's texts as we currently know them. The more we discussed these issues in the seminar, the clearer it became that before we could properly address Galileo's remarks, we needed to develop our own understanding of Aristotle's theory of locomotion. This resulted in an outline for a theory of locomotion with fair capacity to account for a large class of locomotive phenomena, often in mathematical form that Aristotle's own text indicates. Viewed from this theoretical perspective, Galileo's law of free fall remains a highly original contribution, exposing connections between basic Aristotelian parameters that were previously treated as independent within the theory. It is not, however, a principle that requires wholesale abandonment of the Aristotelian theory. In fact, it can be seen as a great new discovery that refines the theory's ability to account for locomotive phenomena.

Development of the ideas that germinated during the joint seminar was aided by discussions with a small group of scholars and students at the Minerva Humanities Center's project on migrating knowledge at Tel Aviv University, headed by Prof. Feldhay, to whom I owe my first gratitude. Without her constant encouragement and support this essay would never have developed beyond my lecture notes and PowerPoint presentations. Particular thanks go to Dr. Michael Elazar, for his endless patience and constructively critical examination of this essay's various aspects. Dr. Ivor Ludlum helped with some difficulties involving Aristotle's Greek, and Appendix B on the threshold of motion in time would not have been possible without Mr. Avi Aroesti's generous help.

Final shaping of the understandings gained during these years into the present essay was made possible by two scholarships at Department 1 of the Max Planck Institute für Wissenschaftsgeschichte in Berlin, under the leadership of Prof. Jürgen Renn, to whom I am indebted for this opportunity. Professor Matteo Valleriani of Department 1 read the entire manuscript, and suggested many useful corrections and improvements. I have also benefitted from Dr. Joyce van Leeuwen's extensive familiarity with the pseudo Aristotelian *Mechanical Problems*. At an advanced stage, select aspects of this work were presented at the Excellence Cluster TOPOI in Berlin. I thank the informed audience I met there for the relevant feedback.

Contents

Introduction

Every product of nature has within itself a source of motion (kinesis) and rest.' Aristotle's philosophy was rooted in nature, especially living nature, and the characteristic of natural beings which called above all for explanation, and offered the greatest challenge to the philosopher, was that they moved about, changed, were born and died. This we have had to notice already, but not its tremendous consequences for his whole conception of the world and its causes.[1]

With these lines, W.K.C. Guthrie opens the chapter titled "Theory of Motion and Theology," in his introductory volume on Aristotle's philosophy. Guthrie dedicates no other chapter in this volume to Aristotle's theory of motion. The chapter's 33 pages focus on prime movers and the fundamentals of cosmological causation. Contrary to the expectations of an unsuspecting modern reader, the discussion says practically nothing about the natural and forced locomotion of inanimate objects, the flight of projectiles, and the effects of the material media in which locomotion takes place. Then, the reader who takes Guthrie's chapter as an initial guide to Aristotle's *Physics* and *On the Heavens*, will be again surprised to find Aristotle devoting a great deal of attention to these subjects. Indeed, if the number of pages dedicated to a specific topic is any indication, then without question locomotion of inanimate objects represents the main subject of specific study in these two works of Aristotle's. Consider the line quoted by Guthrie from Aristotle's *Physics* in slightly expanded context:

"Some things exist, or come into existence, by nature; … and the common feature that characterizes them all seems to be that they have within themselves a principle of movement (or change) and rest—in some cases local only, in others quantitative, as in growth and shrinkage, and in others again qualitative, in the way of modification." (*Physics*, II.i. 192^{b}8-15, Wicksteed and Cornford, 1929)

---

[1]W.K.C. Guthrie, *Aristotle: An Encounter*, (Cambridge: Cambridge University Press, 1981), p. 243. The quoted line from Aristotle is from *Physics*, II.i. 192^{b}13-14.

"Of the things that are, some are by nature, ... and all of them obviously differ from the things not put together by nature. For each of these has in itself a source of motion and rest, either in place, or by growth and shrinkage, or by alteration;" (*Physics*, II.i. 192^{b}8-15, Sachs, 1995)

"Some things exist by nature, ... The obvious difference between all these things and things which are not natural is that each of the natural ones contains within itself a source of change and of stability, in respect of either movement or increase and decrease or alteration." (*Physics*, II.i. 192^{b}8-15, Waterfield, 1996)

Regardless of differences in expression and choice of words, all three translations reveal the distinction that Aristotle gave to locomotion. Wicksteed and Cornford speak of "local" motion, Sachs speaks of motion "in place," Waterfield uses the term "movement" as distinct from growth or alteration. In view of this, Guthrie's nearly total neglect of locomotion under a heading that contains the words "Theory of Motion" may seem somewhat questionable, but the bias is more apparent than real, and this requires a few words of explanation.

Nowadays we associate the word *kinesis* exclusively with change of place, and we see *physics* as a highly mathematized discipline dedicated to the investigation of nature's fundamental forces and the way they make matter move. In Ancient Greek, however, *kinesis* often denotes change in general. More specifically, in Aristotle's extant works it denotes any time-occupying physical process. Aristotle considers the ability to partake in any such process as the distinctive mark of physical objects, and, accordingly, he considers physics as the science of all time-occupying physical processes. To Aristotle, locomotion, namely, change with respect to place or spatial displacement, represents just one type of *kinesis* that physics studies. None of his surviving works develops a systematic theory of locomotion by itself in the manner we have learned to expect from present-day treatises on mechanics. However, the first place allocated to locomotion among types of change in the above quoted paragraph is not accidental. Aristotle sees it as the primary type of change in the sense that the capacity for pure locomotion is a prerequisite to partaking in any other type of change. Locomotion provides the key to identifying the five fundamental material elements in the universe. It also serves Aristotle as a sort of model for all other types of timed change, although he stops short of suggesting that all other types of change must, in the final analysis, be reducible to locomotion. In other words, Aristotle's scope for the subject matter of physics is truly cosmological and engulfs all the sciences we currently refer to as "natural," including medicine. His deep interest in locomotion is beyond reasonable doubt, considering the space devoted to it in *Physics* and *On the Heavens*, but by and large, discussions of locomotion serve in these texts as means to higher ends, and not as an end in itself. Guthrie's chapter on "Theory of Motion and Theology" focuses on these higher ends. Even if this does not justify the near-total neglect of locomotion in the chapter, it does project a fair representation of Aristotle's ultimate aims.

Closely associated with the status of the theory of material locomotion is the use of mathematics in *Physics* and *On the Heavens*. The most superficial perusal of these works will inevitably disclose the crucial importance of mathematical reasoning in them, as well as the formulation of basic principles in mathematical terms. In spite of this, there is much to be said for Sachs's plea to refrain from reading Aristotle's *Physics* as given to mathematical reduction:

> In any encounter with the natural world, it is the kinds of change other than change of place that are most prominent and most productive of wonder. Mathematical physics must erase them all and argue that they were never anything but deceptive appearances of something else, changing in some other way. Why? Because those merely local changes of merely inert bodies can be described mathematically. But if the testimony of the senses has a claim to 'objectivity,' and to be taken seriously, that is at least equal to that of the mathematical imagination, such a reduction is not necessary.[2]

This observation may serve as a useful starting point for reading Aristotle's natural philosophy in general, and in particular *Physics* and *On the Heavens*—the two works that stand at the focus of the present monograph. At the same time, Sachs's view should not tempt anyone to belittle Aristotle's deep and genuine interest in the mathematical theory of locomotion of inanimate objects.[3]

So far, then, two basic features stand out: (1) in *Physics* and *On the Heavens*, the abundance of remarks on the locomotion of inanimate objects and its mathematical representation clearly reveals Aristotle's theoretical interest in the subject; (2) in none of his extant works does Aristotle organize these remarks into a systematic, coherent theory. To assert that he must have had such a theory seems ill advised in the absence of direct evidence to that effect. Equally ill advised, however, is to assert from the silence of the existing records that he did not have such a theory. In other words, while avoiding the anachronistic reading of one's own knowledge into Aristotle's text, one should also take care not to fall into the equally anachronistic reading of one's own ignorance into the text. Setting aside the question of what Aristotle really did or did not have by way of systematic theory, it is still possible to examine how, if at all, his explicit observations on locomotion can be assimilated into a coherent theory. This is what the present study proposes to do without regarding the result as the definitive historical reconstruction of Aristotle's theory of locomotion. At best, historically speaking, it puts forward one possible suggestion of what Aristotle's theory might have been. Most importantly, it shows that without positing any additional basic principles, and without requiring any mathematical means beyond those available to Aristotle, the elements of locomotion in *Physics*

---

[2]J. Sachs, *Aristotle's Physics: A Guided Study* (New Burnswick, Rutgers University Press, 1995), p. 19.

[3]For an illuminating reconstruction of Aristotle's mathematical physics together with a carefully considered defense of the idea that at least with respect to locomotion, Aristotle aspired to a mathematically formulated theory, see Edward Hussey, *Aristotle Physics: Books III and IV*, (Oxford: Clarendon Press, new impression with corrections and additions, 1993), pp. 185–200, and "Aristotle's Mathematical Physics: A Reconstruction," in Lindsay Judson, (ed.), *Aristotle's Physics: A Collection of Essays*, (Oxford: Clarendon Press, 1991), pp. 213–242.

and *On the Heavens* can be integrated, by interpolation only, into a coherent and surprisingly powerful theory of matter in locomotion.

An important guiding idea for this elucidation of Aristotle's theory of locomotion is that the long tradition of Hellenistic and Medieval commentaries on Aristotle must not be read as if motivated by an overriding concern to produce contextualized historical reconstructions of his original ideas. The commentaries may certainly be employed as aids in the attempt to explicate Aristotle's extant texts, provided they are understood as attempts to extend and develop his ideas, not as attempts at historical reconstructions of them of the sort sought by modern historians of science and ideas. To the commentators, Hellenistic, Muslim, and Christian, Aristotle's physics was not an ancient museum piece. Rather, they regarded it as *the* working scientific theory of their day—a theory that espoused great promise together with many frustrating difficulties. They studied Aristotle's physics either to further strengthen and advance it, or to debunk it. Oftentimes, the difficulties that they addressed did not arise from apparent internal inconsistencies, but from attempts to apply Aristotle to cultural and religious contexts that were foreign to his own day and to his way of thinking. Considering this, it seems most useful to try to synthesize Aristotle's theory of locomotion from his available texts directly, without regard to the extrapolated ideas of his commentators. There are many difficulties of squaring various statements in Aristotle's texts, and it is inadvisable to further complicate matters by trying to square his statements with the understanding imposed on them by later students of physical motion. In other words, Aristotle's ideas should not be confounded with those of commentators like Alexander, Simplicius, Philoponus, Ibn Rushd, Aquinas, and Oresme. They strove to stream-line Aristotle into a better philosophy in keeping with the knowledge and beliefs of their own time, and read his works on nature as natural philosophers rather than as historians of ancient scientific ideas. Therefore, a synthesis of Aristotle's thoughts on material bodies in motion independent of his commentators' creative interpretations of it should put us in a better position to understand the commentators' contributions as students of natural philosophy in their own time.

The task of constructing the above suggested synthesis based on Aristotle's texts alone is difficult not only on account of subtleties not easily understood. While such subtleties exist, the primary cause of difficulty is the style of Aristotle's texts. The *Physics* (like *On the Heavens*) is not a didactic exposition of a logical system similar to Euclid's *Elements* or Newton's *Principia*. The latter two begin with definitions, continue with the fundamental postulates to be taken as irreducible, and then proceed to demonstrate theorems and problems in terms of what has been defined and postulated. Aristotle, by contrast, focuses on distinct problems, very often difficulties he identifies in the opinions of his predecessors, and proceeds to analyze and solve them. In the course of such discussions, he does define terms and puts down postulates, but only to the extent that they serve the problem at hand. His own opinions are thus exposed piecemeal, and the reader has the task of collecting the bits and pieces to see how, if at all, they may be synthesized into a coherent theory of material bodies in motion. In general, the *Physics* and *On the Heavens* possess the character of extensive lecture notes designed for advanced learners, assumed to have had prior contact with the basics. Slow, detailed development of basic theory does not go hand in hand with such advanced lectures. The weakness of arguments from silence is, therefore, particularly acute in Aristotle's case. Gaps in argument lines and brief general statements do not necessarily denote the limits of Aristotle's knowledge, and the possibility that his scattered discussions of locomotion reflect aspects of a coherent theory must be considered seriously.

Given these difficulties, the assimilative effort developed in this essay can suggest ways in which Aristotle *might* have thought, but cannot guarantee a reconstruction of what Aristotle *really* thought. Indeed, it has many affinities with the streamlining efforts of Aristotle's past commentators, with one critical difference: Aristotle's theory no longer pretends to the title of leading scientific theory today, and the present study merely wishes to find for it a reasonable streamlining internal logic. This does not in any way exclude appeal to natural phenomena. But the phenomena should be of the type Aristotle appeals to rather than the systematic results of controlled laboratory experiments expected in the framework of modern physics. To the extent that the resulting synthesis reveals paths not taken by Aristotle's Hellenistic, Muslim, and Christian commentators, it could help to bring out their own tacit assumptions and motivations. In so doing it may contribute a step toward a better understanding of the migration and evolution of Aristotle's natural philosophy into various versions of peripatetic natural philosophy.

Finally, a quick note on translation. Classicists quite rightly point out how the idiosyncrasies of English and Greek may cause grotesque misunderstandings of central ideas in translation. Consider the following, from F.M. Cornford's introduction to his translation of Plato's *Republic*, quoted approvingly by Guthrie in *The Greek Philosophers: From Thales to Aristotle*, and disapprovingly by Allan Bloom in the first edition to his own translation of Plato's *Republic*[4]:

> Many key-words, such as 'music', 'gymnastic', 'virtue', 'philosophy', have shifted their meaning or acquired false associations for English ears. One who opened Jowett's version at random and lighted on the statement (at 549 b) that the best guardian for a man's 'virtue' is 'philosophy tempered with music', might run away with the idea that, in order to avoid irregular relations with women, he had better play the violin in the intervals of studying metaphysics. There may be some truth in this; but only after reading widely in other parts of the book would he discover that it was not quite what Plato meant by describing *logos*, combined with *musike*, as the only sure safeguard of *arete*.

One might as well add to this that *Republic* is not the best translation of *Politeia*. The state described in Plato's *Politeia* is not a republic—neither in the Ancient Roman sense of the word, nor in the modern American one—and something like *The Regime* is probably closer to his intention.

---

[4]W.K.C. Guthrie, *The Greek Philosophers: From Thales to Aristotle*, (New York: Harper Tourchbooks, 1975), p. 5. *The Republic of Plato, Translated with Notes and an Interpretive Essay* by Allan Bloom, 2nd Edition (Basic Books, a division of HarperCollins *Publishers*, 1991), pp. xiv–xv.

Since 1929, however, Aristotle's *Physics* has been fully translated into English seven times: Wicksteed and Cornford (1929); Hardie and Gaye (1930), revised in Jonathan Barnes's Oxford Edition of the complete Works of Aristotle (1984, and later reprints); Richard Hope (1961); Hippocrates G. Apostle (1969); Joe Sachs (1995); Robin Waterfield (1996); and Glen Coughlin (2005). Jonathan Barnes's 1984 edition of Aristotle's complete works retains Hardie and Gaye's translation for the most part, but re-renders it significantly at times. Several other translations of select books out of the *Physics* have also been produced in the interim, all of them by well-trained classicists, as keenly aware of the pitfalls of translation as F.M. Cornford. The multitude of professional translations reflects lingering dissatisfactions,[5] and the reasonable conclusion from all this appears to be that in general, no "best" translation of a given paragraph exists. No translation can substitute for going back to the standard Greek text and further back to the original manuscripts at its foundation. However, doing so is very unlikely to yield a final, "once and for all" reading, and exposure to several professional translations of the same paragraph will serve even proficient Greek readers better than yet another new one. Therefore, it seems better suited for the purposes of this monograph to put side by side a sample of the already existing translations of the relevant paragraphs. To preserve continuity of argument, only one of the alternative translations serves the main text, while others may be consulted for comparison in Appendix F.[6] The translations differ, often markedly, in style, and this is particularly noteworthy with regard to Wicksteed and Cornford, who intended their work "to be an interpretative paraphrase" rather than a tight translation.[7] Nevertheless, in the majority of paragraphs quoted here, different translations convey the same idea in different words. In a few exceptional cases, I find some translations preferable to others, and these are noted in the course of the discussion or in an accompanying appendix.

---

[5]See, e.g., J. Sachs, *Aristotle's Physics: A Guided Study*, (New Brunswick: Rutgers University Press, 1995), pp. 4–9, 21–22; Glen Coughlin, *Aristotle: Physics, or Natural Hearing*, pp. xxii–xxx (South Bend, Indiana: St. Augustine's Press, 2005).

[6]To this end, I opted most often for J. Sachs's translation of the *Physics*, not because it is invariably more reliable than others, but because Sachs's simple, direct English appeals to my personal taste and to the way I like to imagine Aristotle's way of thinking. The first two books of *On the Heavens* have benefitted from a new, lucid, and precise translation by Stewart Leggatt (1995), but W.K.C. Guthrie's full translation from 1939 remains an excellent guide, and J.L. Stocks from 1922 (revised in Barnes's 1984 Oxford edition of Aristotle's complete works) is always worth consulting.

[7]Aristotle. 1929. *Physics*. Translated by Wicksteed and Cornford. viii. Cambridge, Mass.: Harvard University Press.

General Plan of the Essay

In the first chapter of this essay, Aristotle's theory of forced locomotion (change of place, or displacement, in time) will be built up as a basic core, followed in the second chapter by a discussion of heaviness, lightness, and Aristotle's concept of natural terrestrial locomotion. Together, the two chapters provide the general theoretical foundation for the ensuing discussions, and its main features are outlined in the summary that follows the second chapter. The third chapter introduces important refinements that clarify and qualify the combined theory of natural and forced motion. For an example of specific application, the fourth chapter shows how analysis of winch action in terms of Aristotle's theory leads to the law of the lever, and compares this analysis to the discussion in problem 1 of the pseudo-Aristotelian *Mechanical Problems*. To further sharpen the unique features of Aristotle's theory of locomotion, the final chapter outlines the in-body impetus theory that Hellenistic and late-antiquity thinkers developed as an alternative to Aristotle's theory.

Important to keep in mind is that Aristotle makes a fundamental distinction between natural and forced motion. By definition, he regards forced motion as unnatural, and as such, forced motion necessarily presupposes natural motion. However, pure natural motion takes place only in the celestial region, which, in Aristotle's geocentric view extends from the moon to the celestial sphere that carries the fixed stars. Celestial motion follows a different set of principles than terrestrial motion, which is the subject of discussion here. In the terrestrial sphere (extending from the center of the universe to the moon), natural motion invariably manifests itself through forced effects. Seeing terrestrial natural motion through these effects requires an understanding of forced motion, which is why this discussion begins with the unnatural and proceeds from there to the natural.

The forced motion core consists of three superimposed levels. They will be referred to as levels 1, 2, and 3. The numbering does not intend to convey any mandatory logical priority, and the order of presentation reflects no more than a descriptive convenience. Logically, they are equally fundamental.

Chapter 1
The Three Levels of Aristotle's Theory of Material Bodies in Forced Motion

Level 1: The Basic Relationship Between Mover, Load, and Speed

The basic aspect of the relationship between mover, load, distance traveled, and time of travel, is in *Physics*, VII.v. While the text introduces the relationship by way of particular examples, its generalized linear mathematical proportionality is unmistakable:

> Since a mover always moves something, in something, and to some extent (and by in something, I mean in time, and by to some extent, I mean that there is some how much of a distance, since always at the same time it is moving something, it also has moved it, so that there will be some so-much through which the thing has been moved, and in so much time), if A is the mover, and B the moved which has been moved as far as the length C in a time as much as D, then in an equal time, an equal power to that of A will move half of B through the double of C, or move it though C in half the time D, for in this way it will be proportional. And if the same power moves the same thing this far in this time, and half the thing that far in half the time, then also half the strength will move half the body an equal distance in an equal time. For example, let E be half the power of A and let F be half the body B; then they stand similarly, and the strength is proportional to the heaviness[1] so that they will move an equal distance in an equal time. (*Physics*, VII.5.249ᵇ27-250ᵃ9, Sachs, 1995)

---

[1] ὁμοίως δὲ ἔχουσι καὶ ἀνάλογον ἡ ἰσχύς πρὸς τὸ βάρος. If Euclid is any indication, then the statement that load and force are proportional (ἀνάλογον) does not suggest a comparison of the ratio of force to load in one case to the ratio of force to load in another. In *Elements*, Book V. def. 3, Euclid defines ratio as follows: Λόγος ἐστὶ δύο μεγεθῶν ὁμογενῶν ἡ κατὰ πηλικότητά ποια σχέσις, "A ratio is a sort of relation in respect of size between magnitudes of the same type." (Heath 1908). Euclid defines proportion in definition 6: Τὰ δὲ τὸν αὐτὸν ἔχοντα λόγον μεγέθη ἀνάλογον καλείσθω, "And let magnitudes having the same ratio be called proportional" (Fitzpatrick 2008). Therefore, to say that things are proportional is to indicate comparison of two ratios, each taken between magnitudes of the same kind. From this point of view, the declared proportionality of force to load means that the ratio of the relevant forces is the same as the ratio of the corresponding loads. In their translation, Hardie and Gaye do not follow this restriction: "then the ratio between the motive power and the weight in the one case is similar and proportionate to the ratio in the other..." From the purely geometrical point of view, lengths of line segments represent all the magnitudes involved. Geometrically speaking, then, Hardie and Gaye's statement works out, since if a:b::c:d, then a:c::b:d, and all the related mathematical manipulations of the proportion are equally legitimate. However, since Aristotle's statement refers to the magnitudes of physically distinct entities, the Euclidean restriction may hold, at least for setting up the basic principle before applying to it all manners of geometrical manipulations.

© The Author(s) 2015
I. Yavetz, *Bodies and Media*, SpringerBriefs in History of Science and Technology, DOI 10.1007/978-3-319-21263-0_1

The mover in this statement is whatever acts directly on the mobile to generate motion. To emphasize this it will be referred to throughout as the effective mover or the effective force, and symbolized by E. The critical importance of this emphasis will only become apparent with the discussion of prolonged projectile motion in level 3. Aristotle illustrates the basic proportions using ratios of 1:2, but other ratios should follow the same pattern. To obtain a general form of Aristotle's principles, let E represent the strength of the effective mover, W the magnitude of the load, S the distance traveled, and T the time of travel. Then the general rule states that for different loads, same effective mover, and same time of travel, the distances are to each other *inversely* as the loads:

$$W_2 : W_1 :: S_1 : S_2|_{T_1=T_2}, \text{ or, in more familiar terms} : \frac{W_2}{W_1} = \frac{S_1}{S_2}\Big|_{T_1=T_2}.$$

Keeping the distance constant, the times are to each other *directly* as the loads:

$$W_1 : W_2 :: T_1 : T_2|_{S_1=S_2}, \text{ or} : \frac{W_1}{W_2} = \frac{T_1}{T_2}\Big|_{S_1=S_2}.$$

Let straight line segments drawn from O represent the times, loads, and distances traveled. Given the loads W_1 and W_2, one distance and one time period, say T_1 and S_1, may be represented by lines of arbitrary length. Construction that satisfies the required proportions then determines the other two magnitudes, T_2 and S_2 (see Fig. 1.1).

Effective mover E, then, makes $\left\{ \begin{array}{l} W_1 \text{ cover } S_1 \text{ in } T_1 \\ W_2 \text{ cover } S_1 \text{ in } T_2 \end{array} \right\}$ such that $T_1 : T_2 :: W_1 : W_2$,

and the same effective mover E makes $\left\{ \begin{array}{l} W_1 \text{ cover } S_1 \text{ in } T_1 \\ W_2 \text{ cover } S_2 \text{ in } T_1 \end{array} \right\}$ such that $S_1 : S_2 :: W_2 : W_1$.

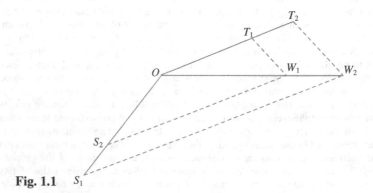

Fig. 1.1 S_1

Therefore, $S_1:S_2::T_2:T_1$. But T_2 is the time in which W_2 covers S_1, and T_1 is the time in which W_2 covers S_2, and since the times are proportional to the distances, it follows that W_2 travels at a uniform speed. Furthermore, since the times in which W_1 and W_2 cover any given distance are as the constant ratio of the loads, the speeds of W_1 and W_2 are to each other in a constant ratio, and since the speed of W_2 is constant, the speed of W_1 must also be constant. Therefore, the first two proportions listed by Aristotle necessarily imply that an effective mover of constant power moves any given load at a constant speed, such that under the same effective mover the speeds are inversely as the loads.

In *On the Heavens*, Aristotle observes that under the same force a smaller load will travel further than a greater one, for: "...as the greater body is to the less, so will be the speed of the lesser body to that of the greater."[2] Therefore, he must have seen that taken together the two statements above are equivalent to requiring the inverse proportionality of speed to load under a constant force. Letting V stand for the speed, then, Aristotle's principle requires:

$$W \propto \frac{T}{S}, \text{or} : W \propto \frac{1}{V}, \text{or} : V \propto \frac{1}{W}.$$

The speed, therefore, turns out to be uniform as long as the effective mover operates on the mobile with constant force. This is explicitly reflected in the quote above: "in an equal time [to D] an equal power to that of A will move half of B through the double of C, or move it though C in half the time D," namely, the motion is such that if in a given time it covers a given distance, then in half the time it covers half the distance. The double ratio once again provides an illustrative example, so generally if in equal time divisions the motion divides the distance into equal segments, then the average speed during any time division is equal to the average speed during the entire time.[3] The conclusion, then, is that for Aristotle, a constant effective force impresses constant speeds in inverse proportion to the loads on which it operates.

All of this, however, holds only for an unchanging effective mover. The opening quote clearly implies the linear effect of changes in the power of the effective mover: if the full effective force moves the full load to a given distance in a given time, then half the effective force will move half the load to the same distance in the same time. Generalizing the linearity without qualification for different effective movers operating on the same load (but see ahead, the threshold of motion):

$$V_1 : V_2 :: E_1 : E_2, \text{or} : \frac{V_1}{V_2} = \frac{E_1}{E_2}.$$

---

[2] τὸ γὰρ τάχος ἕξει τὸ τοῦ ἐλάττονος πρὸς τὸ τοῦ μείζονος ὡς τὸ μεῖζον σῶμα πρὸς τὸ ἔλαττον. *On the Heavens*, III.ii.301^{b}12, (Guthrie 1939).

[3] In other words, if for equal divisions of the total time the motion divides the total distance into segments of equal length, then: $S(t) = n \cdot S(\frac{t}{n}) \Rightarrow \bar{V}(t) \cdot t = n \cdot \bar{V}(\frac{t}{n}) \cdot \frac{t}{n} = \bar{V}(\frac{t}{n}) \cdot t \Rightarrow \bar{V}(t) = \bar{V}(\frac{t}{n})$.

For the same effective mover operating on different loads:

$$V_1 : V_2 :: W_2 : W_1, \text{ or } : \frac{V_1}{V_2} = \frac{W_2}{W_1}.$$

More generally, allowing both the effective force and the load to vary:

$$V \propto \frac{E}{W}. \tag{1.1}$$

Alternatively in explicit compounded form, for two distinct movers E_1, E_2, operating on two distinct loads W_1, W_2, resulting in two corresponding speeds V_1, V_2:

$$\frac{V_1}{V_2} = \frac{E_1}{E_2} \cdot \frac{W_2}{W_1}. \tag{1.2}$$

Characterizing locomotion in terms of speed has explicit support from Aristotle's own writing, and to that extent, the equivalent relations (1.1) and (1.2)[4] are secure paraphrases of his statements. Now, Aristotle considers motion to happen over a period of time, which, no matter how short, is incomparably long with respect to a single instant (see ahead, segments of time, instants of time, and the concept of speed, p. 33). Since he also requires that anything continuous must be indefinitely divisible, then in principle at least, all grades of speed must be available to a body moving at a continuously varying speed. Following Aristotle, then, we may allow a mobile to pass through a continuous spectrum of speed magnitudes.[5] We should still understand with Aristotle that to say of one mobile that it is twice as fast as another means that it would cover the same distance in half the time, or twice the distance at the same time. To be meaningful in Aristotle's way of thinking, however, this statement must have a time segment in mind. The time segments can be as short as we please, and the speed ratio of the two mobiles may change over different time segments. That is to say, to the extent that over some arbitrarily short time segment effective forces E_1 and E_2 operate respectively on loads W_1 and W_2, they respectively impress on the loads speeds V_1 and V_2 according to relation (1.2).

---

[4]Note that compounding proportions as in relation (1.2) avoids the problem of comparing different qualities. The magnitudes obtained on both sides are dimensionless, for we are multiplying the dimensionless ratio of forces by a dimensionless ratio of loads to obtain a third dimensionless magnitude—a ratio of speeds.

[5]Aristotle's recognition of continuous, though non-uniform speed is sufficiently clear in the following where he suggests that motion at a continuously non-uniform speed has a lesser degree of unity than motion at uniform speed: "For the non-uniform is, as it seems, not one, but rather the uniform is one, like the straight, since the non-uniform is divided. ... But sometimes the non-uniformity is not in that which moves, nor in the when, nor in the direction, but in the way it moves. For it is sometimes distinguished in quickness and slowness, and that of which the speed is the same is uniform, that of which it is not, non-uniform. Hence speed and slowness are not kinds of motion, nor specific differences, because they go along with every specific difference of kind. ... So the non-uniform motion is one by being continuous, but less so," (*Physics*, V.iv.228[b]17-229[a]5, Sachs).

For the same arbitrarily short time segment, V_1 and V_2 are to each other as the distance segments S_1 and S_2 that each of the loads covered, respectively. For all practical purposes, this should suffice, with partial justification for applying this to continuously variable forces and speeds coming from the "method of exhaustion," for which Aristotle's older contemporary Eudoxos of Cnidos had laid the foundations, and which Euclid employs in Book 12 of his *Elements*.[6]

As a matter of principle, however, Aristotle does not have our current mathematical concept of "instantaneous speed" as the limiting ratio between distance traveled and a time segment as the time segment becomes indefinitely short. To work sensibly, this notion needs to meet two requirements: a general definition of "limit," and a specific proof that the limit as generally defined exists for the case under consideration. I could find no indication in Aristotle's extant works that he thought along such lines. Without that, it seems difficult to square the notion of instantaneous speed with Aristotle's idea of motion as something that makes no sense at a single instant of time. None of this undermines the conception of continuously varying grades of speed in the context of relationship (1.2) above: given that effective force E moves load W_1 at the speed V, we can calculate the speed at which the same force would move load W_2. However, at least to a modern reader, this does leave a theoretical ambiguity when attempting to switch from talking in terms of speed to talking in terms of distances covered in time segments and vise-versa.

Most important is never to mistake relationships (1.1) and (1.2) between speed, load, and effective force for an idealized law of motion without resistance. There is no vacuum for Aristotle, and no motion without interaction with a surrounding medium. A vacuous space, to Aristotle, does not produce zero resistance to motion. It produces inconsistent results and irresolvable paradoxes. For Aristotle, the vacuum is really and truly a counterfactual—a natural impossibility.[7]

---

[6]In Book 12 proposition 2, for example, Euclid proves that the areas of two given circles are to each other as the squares of their diameters. The procedure involves the comparison of polygons with progressively greater number of sides (beginning with squares) inscribed and circumscribed in and about one of the circles. This is done for the purpose of showing that any area whose ratio to the area of the first circle either exceeds or falls below the ratio of the squares of the two diameters also exceeds or falls below the area of the second circle. Therefore assuming the contrary of the proposition invariably leads to a contradiction. In other words, Euclid demonstrates a theorem on the continuously curved circle by an argument on straight segments of polygons with arbitrarily many sides. This line of argument is called the method of exhaustion because it always shows by similar reduction to absurd that the difference between two magnitudes can be "exhausted," namely, made smaller than any finite magnitude, and hence the two magnitudes must be equal. However, continuous variability can take many forms other than circles, and therefore the general move from fixed segments of time and distance to continuously varying speed requires a formally generalized method of exhaustion. Euclid provides no such formalization and neither does Archimedes, who applied exhaustive techniques to problems not studied in Euclid's elements.

[7]This may be contrasted with Galileo's law of free fall in a vacuum. For Galileo, the vacuum is a real natural state, perhaps difficult to obtain practically, but it is not in principle counterfactual, nor also is his law of free fall: to the extent that observations in fact appear to contradict it, this is only because the observations were not carried out in a vacuum, as they should have been. Aristotle would throw up his hands in despair at the very notion of observing motion in a vacuum.

Therefore, the two relationships must be taken to reveal only a partial aspect of the full relationship between motion and moving cause. Any pretense to draw from them conclusions applicable to real physical motion is unadvisable, for it fails to consider other ever-present aspects that cannot be idealized out of the picture. Essential for completing the account is a consideration of the medium, and it adds two further levels: resistance to motion, and prolongation of motion.

Level 2: The Medium and Resistance to Motion

That through which it is carried is a cause which impedes it most when *it* is carried in the opposite direction, but secondly also when it is still, and more so that which is not easy to divide, and this is what is thicker. So A will have been carried through B in the time C, but through D, which is thinner, in the time E, if the distance through B is equal to that through D, the times being in the same ratio as the impeding bodies. For let B be water and D air; by as much as air is thinner and less bodily than water, by this much will A have been carried faster through D than through B. Let the speed have to the speed the same ratio that stands between air and water. So if the first is doubly thin, it will go through B in a time double that through D, and time C will be the double of E. And always, by so much as that through which it is carried is less bodily, less resistant, and more easily divided, the faster will it have been carried. (*Physics*, IV.8.215^{a}30-215^{b}12, Sachs, 1995).

The first sentence about the greater resistance offered when the medium moves against the moving body is notable for its suggestion that the medium's resistance grows with flow-speed. However, before making confident use of this suggestion, one would wish for Aristotle to have compared the push experienced by a stationary body in a medium that flows around it to the push experienced by the body when the medium is stationary and the body moves through it at the same speed in the opposite direction. For the ensuing discussion we put off this issue, and proceed with the case of objects moving through stationary media.

To begin with, then, we note that the medium's resistance manifests itself actively only in the presence of motion. Therefore, the formulated relationship between speed and resistance provides a partial aspect only, for, in the case of forced locomotion, without an effective mover that operates on a mobile to move it, the medium will have nothing to resist. This means that the explicit mathematical relations given in the text do not exhaust the subject, for they do not relate the resistance to the other effects on speed, such as the magnitudes of the effective mover and the load. Keeping this in mind, consider first the explicitly stated mathematical relations. They state that the times, T_1, T_2 taken to move through a given distance in two media, are to each other as the resistances ρ_1 and ρ_2 offered by the media. Alternatively, we may relate the resistances to the speeds rather than the times, as the text explicitly suggests:

$$V_1 : V_2 :: \rho_2 : \rho_1, \text{ or} : \frac{V_1}{V_2} = \frac{\rho_2}{\rho_1}.$$

Either way, this must tacitly assume "all other things being equal," namely, the exact same mobile, moved by the exact same effective force. Furthermore, if the mobile's shape is irregular, then its orientation relative to the direction of motion must also be the same, for Aristotle is clearly aware that shape affects the resistance to motion through a medium (*On the Heavens*, IV.6.313^{a}14-15). Under these restrictions, ρ is a characteristic of the medium itself, independent of what moves in it and the speed of motion. Generally, however, ρ differs for different mobiles in the same medium and under the same effective force, and should be considered to reflect a mutual interaction influenced by both mobile and medium.

Nowhere in the extant works of Aristotle is there instruction on how to combine this isolated observation on resistance and motion with the basic relationship represented either by (1.1) or (1.2) in level 1. Indeed, the observations on the effects of medium resistance are given in Book 4, so they precede in order of presentation the relations formulated in Book 7. It may be that no combining instruction exists because Aristotle did not know how to combine the effect on speed of medium resistance with the effects of effective mover and load. However, to assume this without some documented indication of failure (such as an explicit admission by Aristotle to that effect, or a report to that effect by his followers), would be no less speculative than to suggest that he did know how to do this, but did not see fit to formulate it in the context of his discussions (see the general introductory remark on the style of Aristotle's texts). The object of *Physics* IV is not to expose systematically how moving bodies interact with the media that they move in, but rather to reveal the paradoxical nature of positing the existence of void. The discussion of speed and medium resistance is accordingly limited only to what the argument at hand requires.

Having said that, it is not particularly difficult to fit the present observations on medium resistance into the framework suggested by level 1. The latter teaches that speed is directly proportional to the effective mover's strength and inversely proportional to the magnitude of the load. Compound into this the inverse proportionality of the speed and the medium's resistance, and then, if load W_1 is moved by effective mover E_1 through medium ρ_1, while load W_2 is moved by effective mover E_2 through medium ρ_2, then (ρ_1 and ρ_2 stand for the respective media's resistance to being cleaved by the mobiles):

$$\frac{V_1}{V_2} = \frac{E_1}{E_2} \cdot \frac{W_2 \cdot \rho_2}{W_1 \cdot \rho_1}, \text{ or, alternatively} : \ V \propto \frac{E}{W \cdot \rho} \tag{1.3}$$

To the extent that the motions under examination take place in the same medium, and the loads, W_1 and W_2, have the exact same shape, size, and spatial orientation relative to the direction of travel, relation (1.3) here reduces to relation (1.2) of level 1. So consideration of the medium's resistance provides a practical qualification under which relation (1.2) may be applied not as an unrealistic ideal, but as a practical principle. Simple and intuitive as this may seem, we are about to learn that taking the medium into account with relation (1.3) still falls short of representing Aristotle's full theory of matter in motion.

Level 3: Why Motion Persists After the Initial Moving Cause Has Been Removed

But about things that change place, it would be good first to raise a certain impasse.[8] For if everything that moves is moved by something, how is it that some of those that do not themselves move themselves are moved continuously when they are not touching the mover, such as things that are thrown? And if the mover at the same time also causes something else to move, such as the air, which when it is moved causes motion, it is similarly impossible, once *it* is not touching the thing that first moved it, for it to be moved either, but all of them must be moved at the same time and have stopped whenever the first mover stops, even if it, as does the magnet, makes what it moves able to cause motion. But it is necessary to say this, that the thing first causing the motion makes the air or the water or some other thing that is naturally such as to be able to cause motion as well as be *be* able to cause motion; but it does not stop causing motion at the same time it stops being moved, but stops being moved at the same time that its mover stops moving it, and yet is still a cause of motion. And for this reason it moves something else next to it; and it is the same story with this. But it comes to an end when a lesser power of causing motion continually becomes present in the next thing. And that comes to a final stop when the preceding mover no longer makes it able to cause motion, but only makes it be moved. And these must stop at the same time, the mover and the moved, as well as the whole motion. So this sort of motion comes about in things that admit of something being moved and something being at rest, and is not continuous, though it seems to be, since it belongs to things that are either in a row or touching, for the thing causing the motion is not one, but a series of things next to each other. This is why such a motion appears in air or water, the kind of motion that some people say is circular replacement. But it is impossible to resolve the things raised as impasses except in the way described.[9] Circular replacement makes everything be moved and cause motion at the same time, and so also stop at the same time; but the present example appears to be some one thing being moved continuously. By what,

---

[8]Aristotle presents the difficulty under discussion as specific to locomotion (he confines the problem to carried things, or things being borne: Περὶ δὲ τῶν φερομένων). This should give pause to anyone who expects to find in Aristotle's theory of locomotion a one-to-one analogy with other types of change.

[9]The care Aristotle took to differentiate his theory from the "mutual replacement" that he designates by the term "antiperistasis," shows that he did not originate the need to explain why objects continue to move without apparent contact with a mover. Plato's *Timaeus*, 57E, expresses this concern quite clearly: "For it is difficult, or rather impossible, that what is to be moved should exist without that which is to move it, or what is to cause motion without that which is to be moved by it." (Cornford 1937, p. 240). Plutarch, in Platonic Questions, 7.5, suggests that the air that closes in behind a moving object pushes it forward, and this is usually taken to represent the theory of "antiperistasis" that Aristotle rejects. The origin of this notion is unclear, and cannot confidently be attributed to Plato. No extant Platonic work offers a systematic discussion concerning the prolonged motion of projectiles, once separated from the projector. In *Timaeus* 79–80 Plato rejects the possibility of a vacuum, and hence follows the conclusion that all flows must close back on themselves, be they generated by breathing or by objects rolling on the ground or by objects projected through the air. However, nowhere does the *Timaeus* claim that the mutual pushes that accompany such closed flow patterns creates a net push that prolongs the motion. Further investigation into presocratic concepts of prolonged projectile motion is beyond the scope of our discussion. The idea that what requires explanation is not why bodies continue to move, but rather why they stop once set in motion, began to guide physical thinking productively only after Galileo's work.

then? For it is not moved continuously by the same thing." (*Physics* VIII.x.266ᵇ28-267ᵃ21, Sachs, 1995).

The first thing to note is that Aristotle takes great care to dissociate the active locomotion of media like air and water from their function as effective movers. An original agent of motion loads the medium with moving power, *in addition* to setting it in motion, and not as *a consequence* of setting it in motion:

> But it is necessary to say this, that the thing first causing the motion makes the air or the water or some other thing that is naturally such as to be able to cause motion as well as be moved *be* able to cause motion; but it does not stop causing motion at the same time it stops being moved, but stops being moved at the same time that its mover stops moving it, and yet is still a cause of motion. (Sachs)

How an original mover impregnates the medium with moving power remains a complete mystery. Even a partial account of this requires explicit answers to questions such as how will activates sinews and limbs; why a flexed muscle can effect motion while a relaxed one cannot; what is the nature of the power in a drawn bowstring. Aristotle does not provide answers to such questions. For the sake of streamlining his theory of locomotion it suffices to take it as a basic postulate that original movers impregnate the medium with direct moving power.

Aristotle stresses that the sustainment of motion involves different medium parts at different times and that consecutive parts of the medium take up the action progressively. Moving ability is transferred from one medium part to the next as the mobile moves through them, while each, in turn, acts back on the mobile to move it further, resulting with repeated transfers of moving ability to contiguous medium parts as the motion progresses. To the extent that motion through the medium takes time, so does the transmission of action.[10] In the course of these exchanges, consecutive medium parts always receive a fraction of their predecessors' power. Therefore, the effective moving force that operates on the mobile decays in time. Now, Aristotle considers the medium to be continuous in the sense that it is divisible without limit. It follows from this spatial continuity of the medium that just as the motion of the mobile through the medium is continuous, the mobile experiences the medium-imposed effective force as continuously decaying in time (but supplied by consecutive parts with respect to the identity of the effective mover).[11]

While the text describes the force-holding capacity as typical of media such as air and water, it should be clear that motion without any medium is as impossible

---

[10]Similarly, in *On Prophesying by Dreams*, Chapter II.464ᵃ6-9, Aristotle describes disturbances that travel through air and water, to affect distant things without the presence of the original mover. Under Aristotle's dynamics, with speed proportional to force, this, too, could not have been a time consuming process without separating the motion causing ability of the medium from its active motion.

[11]In *Physics* VI.i.231ᵃ21-23 Aristotle defines "as 'continuous' things whose limits formed a unity, as 'in contact' things whose limits are together, and as 'successive' things which have nothing of the same kind as themselves between them." (Waterfield) In the account of persisting motion, Aristotle describes the consecutive air elements as either touching (in contact) or successive, but not continuous.

for Aristotle as the impossibility of vacuum. Therefore, the aspect of motion addressed here is not secondary. It is no less primary than the relationships formulated in levels 1 and 2.

Contrary to Aristotle's discussions of effective mover, load, resistance, and speed in levels 1 and 2, no mathematical relationships are suggested in level 3. The explicit mathematical cast of levels 1 and 2 rules out the possibility that Aristotle considered locomotion as essentially not given to mathematization. Beyond this, it is impossible to tell whether the absence of mathematical casting for prolonged projectile motion reflects Aristotle's unsuccessful attempts to do so, or his reluctance to further complicate this long digression from the chapter's main argument with the technical formalism of proportions.[12]

This is about as much as can be said with plausible confidence about Aristotle's proposed account of prolonged projectile motion. The text says nothing about how to join this new feature to the other aspects connecting movers, loads, medium resistance, and speed of motion. For all we know, Aristotle may not have succeeded to effect a coherent synthesis of all these elements. Indeed, he may have had no interest in producing such a synthesis, so that discussing it in terms of success and failure is itself historically prejudiced. But then, the lecture notes collected into the works we now possess outline discussions intended not for novices but for initiates, ready to discuss advanced problems relating to the opinions of other thinkers. It is, therefore, of historiographical if not historical importance to see whether or not the elements outlined so far can, indeed, be synthesized into a coherent theory of matter in motion. This may be done with the aid of the following propositions:

1. The medium does not possess moving ability in and of itself, and so it must be that as the original mover engages the mobile, it also infuses moving ability into the medium.
2. As pushing or pulling of a heavy load begins, its speed does not jump from 0 to some finite speed proportional to the exertion of the original mover. Rather, the speed develops gradually from rest, as does the effective force applied by the medium. To keep this in line with the rules of level 1, let the force in the medium always do the effective moving, while the original mover serves only to impress moving force on the medium. The importance of this cannot be overstated: all the relationships between effective force, load, resistance, and speed in levels 1 and 2 invariably refer to the moving force in the medium, not to the force of the original mover (e.g. the degree of exertion that a horse applies to move a load). The original mover, say, the horse that strains against a heavy wagon, does not directly move it, but infuses the surrounding medium with the effective force that does the actual moving. The opposing stresses against the effective force that arise from load and resistance reflect back to the original

---

[12]The purpose of the discussion in *Physics* VIII.x is not to introduce systematically the theory of prolonged projectile motion, but to list all the explanations of motion that could not account for the eternal and unceasing motions in the celestial region. The end result is that primary unmoved movers must exist.

mover creating a sense of effort, but the relations formulated in levels 1 and 2 do not tell us how exactly the original mover relates to E and the other characteristics of motion.[13]

3. In keeping with the strict linearity of all the relationships explicitly stated by Aristotle, it seems most reasonable to assume that the rate at which the medium receives force is directly proportional to the strength of the original mover.

4. The rate of force loss in the medium must also be determined somehow, and here some clues may be obtained from everyday phenomena. A water container with a hole at its bottom loses water more rapidly the more water it contains. Flexible balloons lose water or air through an opening the more water or air they contain. Heated objects cool down more rapidly the hotter they are. Similarly, the rate of force-loss in the medium should be taken as directly proportional to the amount of force already in it. In other words, the net rate of force accumulation in the medium is directly proportional to the difference between the force of the original mover and the force already stored in the medium. This will ensure that the force in the medium can never exceed the force of its source—the original mover.

5. The decay of force in the medium should be taken as an irreducible property of the medium, independent of its resistance to being cleaved.

Full mathematical synthesis of these elements with the principles of levels 1 and 2 requires a systematic mathematical kinetics of variable speeds. Since nothing of this sort may be found in Aristotle's extant works, his overall theory of matter in motion cannot be expressed using only the mathematical terms that he explicitly supplies. Furthermore, the discussion in *Physics* VIII.10 addresses only the decay of motion once mover and mobile disconnect. Surely, however, if moving force transfers through consecutive medium layers as the mobile moves, it must also do so while the original mover is still in contact with the mobile.

The theory could work along the following qualitative lines: When a mover first applies itself to a mobile, it begins to load the surrounding medium[14] with moving force (indirectly, across the surface of contact between mobile and medium). Aristotle stipulates that the medium does not operate back on the mobile with all the

---

[13]Collisions represent a particularly vexing difficulty, and will be dealt with separately. At this stage it suffices to note that a collision between projectiles involves a meeting of two already operating medium forces, with no further involvement of original movers. Therefore, movement after the collision must reflect the properly computed combination of the two medium forces prior to the collision.

[14]It may be observed that the force-holding medium layer must be very thin, since one cannot move an object by slowly gliding a hand one tenth of a millimeter over it. However, the medium layer must have a finite depth, for a mere surface is not "body" in Aristotle's sense. This would require a distinction between direct physical contact and moving contact. Aristotle never raises these issues, whether he thought of them or not. The volume and shape of the medium participating in this process is not addressed, and we proceed with Aristotle just to assume that some volume of medium, probably a thin layer of unknown shape and size, participates in this motion-prolonging interaction.

force transferred into it, and so we assume that at any given moment the force in the medium suffers a fractional decay. In the beginning there is no force in the medium, hence none decays, and the loading rate far exceeds the rate of decay. Therefore, at this stage of contact between the original mover and the mobile the effective moving force grows rapidly, the speed of motion must accelerate, and if it began from rest, it must accelerate continuously from rest through all grades of speed. It is false, then, to conclude from *Physics* VII.5 that once an external mover starting from rest applies itself to a load, it induces a motion that switches abruptly from rest to a finite speed proportional to the ratio of force to load. Rather, since all motion takes place in media, it is the discussion of prolonged projectile motion that teaches how properly to apply the principles of *Physics* VII.5 (level 1) to the actual generation of motion by effective movers.

In other words, it is not necessary to force Aristotle's theory into overt conflict with simple phenomena as long as it suggests a way to accommodate them. One way of achieving this is by allowing the medium always to act as the effective mover, while the original mover becomes the source that begins and supports the gradual infusion of the medium with moving force as motion progresses.[15] As the effective moving force of the medium grows, so does the amount of force operating on the mobile to speed it up, but so also grows the rate of force disintegration in the medium. Eventually a time will come when the rate of force disintegration in the medium matches the rate at which the original mover infuses force into the medium.[16] From that time on there will be no further increase in effective moving force, the speed will no longer change, and remain constant as long as the mover is in contact with the mobile (provided the mover itself works with constant force). Once the original mover breaks contact with the mobile, nothing compensates for the continuing loss of moving force in the medium, and hence the speed will diminish continuously until no further moving force remains.

The idea that all locomotion is effected by the medium surely seems strange to anyone accustomed to post-Newtonian dynamics. It requires that when one moves a hand through the air, or when one strolls down the street, it is the surrounding air that effects the movement directly, not the arm-muscles or the push of the feet against the ground. Why, however, once the throwing arm's muscles relax, does the entire arm still continue to swing, in what pitchers and stone throwers refer to as "follow through"? When sprinters cross the finish line, why do they stress their

---

[15]One may object that nothing of the sort is indicated in *Physics* VII.5, but then, the medium is never mentioned there either, while its ever-present critical role is beyond plausible dispute. Furthermore, the objection flies in the face of Aristotle's explicit statement in *On the Heavens* III.2.301^{b}20-30 that in all motion, natural or forced, the medium acts as an intermediary without which no forced motion would be possible (see ahead, next section).

[16]This relates to the overall net effect on the mobile. Actually it should be considered that consecutive medium shells progressively receive moving force from their preceding medium shells, plus the constant contribution of force from the original mover. With the principle that force-loss in the medium is proportional to the force held and that the medium is infinitely divisible this will lead to the overall effect on the mobile as described above.

muscles against the body's motion in order to stop quickly? In the absence of inertia, their own motion, which they now strain to stop, must be effected by some invisible agent. Aristotle locates this agent in the medium. The extra effort is spent to deplete the medium of its driving force faster than the rate at which this force decays on its own. In this manner Aristotle's active medium accounts for a wide range of similar phenomena that we attribute to a concept of inertia that makes sense only with the association of force with acceleration, and this Newtonian inertia should not be confused with what became known in the middle ages as "impetus." In the next section we shall see that according to Aristotle, neither natural nor forced motion can take place without a medium. This means that the medium effects motion not only once the original mover becomes detached from the mobile, but also throughout the period of contact between them. Given all this, attributing independent dislocating ability to the original mover amounts to an unnecessary redundancy that the assumption in proposition 2 above avoids.

The attempted counter argument that we do not feel the push or pull of the medium is misdirected. The medium expresses itself not as a pusher or puller. In particular, the notion that in Aristotle's theory the medium prolongs motion with some sort of wind action is completely out of place.[17] Pushes and pulls are expressions of external original movers or of external obstacles to motion. As the effective mover, the medium expresses itself to our senses through the effort that *we* apply to infuse it with force either to effect motion or to stop an already existing motion. Only the medium's resistance to being cleaved is sensed as an external counter-push.[18]

Throughout this, it is imperative not to think of the effective force as something that flows from the medium into the mobile in order to move it. The medium's force is the immediate cause of a mobile's motion. Weakening is not the result of force leaving the medium and entering the mobile. The special trait of the medium is that when some original mover acts on it, it acquires a measure of moving force that can, in turn, apply to move a mobile, but never to move the same medium element itself. The flow of force in this process is therefore strictly unidirectional—into the medium, via the mobiles that move through it. The reduction in the medium's moving force is, therefore, the result of gradual reduction of force as moving ability migrates through the medium while the mobile moves. This seemingly over-complicated process is Aristotle's way to avoid postulating inanimate self-movers.[19] Inanimate,

---

[17]Indeed, the sustainment of motion in a breath of air once the blower has ceased to blow needs to be explained as coming about from air layers being moved by other air layers while also obtaining from them moving power that they can transfer further to neighboring air elements.

[18]Wind, therefore, is itself an original mover with its own strength. It moves by colliding with and pushing other objects, and this is possible only because of the medium's other, independent property of resistance to being cleaved. Even a very fast wind cannot hurl a stone like a well-trained pitcher, showing that as an original mover, the wind is rather weak—it can load only a small amount of moving force into the medium layer around a stone that acts as its effective mover.

[19]A single homogeneous object cannot move itself according to Aristotle (*Physics*, VIII, iv.255^{a}12-15). Aristotle does consider animals to be self-movers, but only because they are compound and not simple, and embrace within their physical extension a moving agent that differs from what it moves. In this case, the animal's soul causes the sinews to strain against the limb

soulless things cannot move themselves, according to Aristotle. Therefore, wherever an inanimate mobile moves, an external mover must be associated with the motion.[20]

So far then, we have a partially quantified theory of how, while direct contact with an original mover persists, a mobile's motion in a material medium accelerates to a terminal speed, and how the motion decelerates once contact with the original mover is broken. However, this is still an incomplete account of Aristotle's theory of matter in motion, because his discussion of natural motion, in particular the natural fall of heavy objects reveals additional, unexpected aspects. Furthermore, in the context of discussing natural and forced motion, Aristotle provides the most explicit statement to the effect that all motion, without exception, is produced through the mediation of the surrounding medium.

Appendix C shows that it is possible to turn the generally qualitative account of level 3 into a consistent mathematical theory that incorporates the relations of levels 1 and 2 between effective force, load, medium resistance and speed. It is practically impossible to imagine Aristotle or any of his successors producing such a mathematical formulation, considering that it is cast in terms of first order linear differential equations. Its only usefulness is in showing that Aristotle's scattered notes on matter in motion may, in principle, be synthesized into an internally consistent theory. No historical conclusions regarding proper interpretations of Aristotle's extant texts can be based on it.

(Footnote 19 continued)

whose movement relocates the entire body, which carries the soul within it. This is analogous to the boatman who strains against the oars, in order to move the boat, which carries the boatman (*Physics*, VIII.iv.254[b]13-35). And all of these stresses and strains are internal to the original mover. In themselves, they do not suffice to explain the active spatial displacement that follows as a result. To suggest this is to ignore Aristotle's explicit discussions of the medium's active role as the intermediary that brings about active motion. In the end, the straining original mover always effects motion indirectly by infusing the surrounding medium with effective moving power, and activating it to become the direct instrument of the motion.

[20]Aristotle's unwavering objection to the notion that voids can exist in nature can now be further illuminated. Active locomotion requires a definite ratio between the medium's coefficients of resistance and force disintegration. Different media have different ratios of resistance to disintegration, and the two magnitudes are mutually independent—one cannot be computed as a function of the other. Now a void neither resists being cleaved, nor holds a gradually disintegrating force, so the coefficients of resistance and disintegration must both be zero. As different media rarify and become less substantial they may seem to approach the state of void, but the mutual independence of resistance and disintegration means that there is no way to fix some limiting value for the ratio of the two as they tend to zero while different media become progressively rare. Reification of material media cannot, therefore, lead by a limiting process to the specification of the void's ratio of capacity and leakage, leaving the void with an unspecifiable ratio of zero to zero. Motion in a void cannot, therefore, be uniquely and unequivocally specified. It is indefinable, and cannot therefore come into active existence. Aristotle insists that something that cannot in principle come into active existence, cannot exist potentially either. Motion and void are, therefore, mutually incompatible, and since the nature of a thing is fundamentally defined by Aristotle as the manner of its characteristic movement, the void has no nature—it is fundamentally a natural impossibility.

Chapter 2
Heaviness, Lightness, Sinking and Floating

Weight, for Aristotle, is inextricably bound with matter and motion in a way that goes to the very core of his concepts of nature (*physis*), the student of nature (physicist), and the accompanying body of knowledge, namely, physics. Motion in this context refers to anything that changes in time. This means all forms of change, excepting sudden transitions. The transition from absence to presence of contact, for example, falls outside of what Aristotle calls motion and outside of study and analysis as physical process. Things either do or do not make contact, and cannot be in half or quarter contact. Once established, contact may weaken or intensify in time, and these are proper motions. But if contact weakens gradually to the point of breaking, then the first instant of no contact belongs to the next phase of timed change, which is now in the category of drawing apart, and no longer in the category of weakening contact. So only continuous change comes under the domain of physical process, where continuous means infinitely divisible in time. With this qualification, the study of physical process in Aristotle's sense, namely of timed transitions, encompasses all the natural sciences, and not just the specialized subject that nowadays goes by the name of physics.

All material objects are capable of change, according to Aristotle, and whatever changes through time must have a material substrate.[1] While a great variety of

---

[1]The question of what is the essential material substrate is not relevant to our discussion, but the following may be noted: Pure substances like gold or silver may be taken for the material substrate of statues, but the differences between gold and silver show that each of them is already a combination of material substrate and form. To avoid endless regression, Aristotle posits the existence of a formless fundamental substrate that can carry any and all forms relating to matter. *Hulé*—the term associated with the fundamental substrate is not exclusive to it. Aristotle would use *hulé* as the golden substrate of a statue in one context, and as the fundamental substrate, that carries the form of a material element, like earth, in another. In the latter sense, even the five elements are not fundamental, but *hulé* can never actively exist in pure formless state. Pure *hulé*, then, has no independent existence any more than pure form. The following restriction, however, must still always hold: all the *hulé* in the eternal and never created cosmos is, and always has been, locked into the celestial and terrestrial elements. The ether is non-transmutable, and while the terrestrial elements are mutually transmutable, none can appear out of nothing, and none can disappear into nothing. Therefore, the finite size of the cosmos necessarily implies a limited supply of *hulé*, and the idea that *hulé* is potentially of unlimited quantity cannot be defended against the eternity and strict finiteness of the Aristotelian universe.

© The Author(s) 2015
I. Yavetz, *Bodies and Media*, SpringerBriefs in History of Science and Technology, DOI 10.1007/978-3-319-21263-0_2

transitions is available to material objects, they all share the ability to partake in one fundamental type of motion, namely, spatial displacement, or locomotion. Since nothing is more fundamental than motion that involves displacement accompanied by no other change, the fundamental elements of motion must belong to the category of pure locomotion. To the extent that *material* elements exist, they must exhibit motions that serve as first principles of motion in the sense that "...first principles must not be derived from one another nor from anything else, while everything has to be derived from them." (*Physics* I.v.188ᵃ27-28). It must be stressed that Aristotle does not suggest that all natural motions, such as color change, heating and cooling, or hardening and softening, are reducible to locomotion. He sees locomotion as fundamental in the sense that anything capable of some form of change must also be capable of locomotion.[2] In that sense locomotion is fundamental to all motion. Therefore, the first principles of locomotion, to which all locomotion is reducible, are indicative of the fundamental elements of matter, which must be present in all that changes. Since locomotion takes place along paths:

> All physical bodies and magnitudes are in themselves, we say, mobile in respect of place; for we maintain that nature is a principle of movement in them. And all movement in respect of place, which we call locomotion, is either straight, in a circle, or a combination of these; for these two alone are simple. The reason is that these magnitudes alone are simple, the straight line and the circular. Movement about the center, then, is in a circle, movement upwards and downwards is rectilinear. By 'movement upwards' I mean movement away from the centre, by 'movement downwards' that towards the centre. So that all simple locomotion must be away from the centre, towards the centre, or about the centre. (*On the Heavens*, I.ii.268ᵇ14-24, Leggatt 1995).

Note that Aristotle refers to the line and the circle as magnitudes, not merely as qualitative forms. A given line is inalienable from its length, just as a circle cannot be imagined without a size for its radius. To Aristotle, the study of physics already at its most elementary level must involve magnitudes. Merely qualitative considerations can never fully capture the subject matter of physics.

> Since some bodies are simple and others compounds of these (I mean by 'simple' all those that have a principle of movement according to nature, such as fire, earth, their forms, and

---

[2]"A thing is said to be prior when its existence is a prerequisite for the other things to exist, but not vice versa, [...] So Change of place must be primary, because neither increase nor decrease nor alteration, nor again coming to be or ceasing to be, are necessary prerequisites for movement, but the continuous movement which the first agent of change imparts is a necessary prerequisite for the existence of these other kinds of change." (*Physics*, VIII.vii. 260ᵇ17–18 [...] 26–28, Waterfield). The reference to a prime agent of movement is explicit here, and the intention must be to the prime movers studied in the *Metaphysics*. These agents induce the uniform rotations of the celestial ether spheres, which force continuous locomotion in the terrestrial region, as described in *Meteorology* I. ii.339ᵃ20–32. But, "...there is no necessity at all for the thing that changes place to be either increased or altered, nor, certainly, to come into being or be destroyed; but none of these is possible if there is not the continuous process which the first mover sets in motion." (*Physics*, VIII. vii.260ᵇ28-29, Sachs) All of this shows that locomotion is a necessary (though not in itself sufficient) condition for all other types of change.

their congeners), movements must also be simple or some kind of combination, and simple bodies must have simple movements, compound bodies combined movements (moving according to that component which predominates).

Thus, if there is such a thing as simple movement, and movement in a circle is simple, and the movement of a simple body is simple and simple movement belongs to a simple body (for even if simple movement belongs to a compound body, it will belong according to that component which predominates), there must be a simple body that is such as to move in a circle according to its own nature. (*On the Heavens*, I.ii.268^{b}26-269^{a}6, Leggatt 1995).

Guided by these observations, Aristotle identifies three types of simple motions, and hence three types of elemental matter: straight upward motion exhibits the nature of the elementally light; straight downward motion exhibits the nature of the elementally heavy; circular motion about the center exhibits the nature of the elementally weightless. By upward and downward in this context, Aristotle always means respectively away from or toward the one and only center of the universe, so the entire discussion presupposes a spherical geometry for the entire material universe.

Since some of the things that have been said are being assumed, while others have been proved, it is clear that all body does not possess lightness or weight. What we mean by 'the heavy' and 'the light' must be assumed at the moment sufficiently for our present need, but more accurately later, when we shall examine their essence. Thus, let that which is such as to move towards the centre be 'heavy', that such as to move from the centre 'light', that which sinks below all downward-moving bodies 'heaviest', and that which rises above all upward-moving bodies 'lightest'. Then everything that moves down or up has to possess lightness, weight, or both (this last, however, not in relation to the same thing; for bodies are heavy and light in relation to one another, as is air in relation to water, and water to earth). But the body that moves in a circle cannot possess weight or lightness; for it cannot move towards or from the centre either naturally or counter-naturally. (*On the Heavens*, I. iii,269^{b}18-31, Leggatt 1995).

A word on nomenclature is required here. Translations of Aristotle often use the words "weight" and "heaviness" interchangeably for the Greek "*baros*." This could result in confusing relative and absolute weightlessness. That a proper mixture of cork and steel neither sinks nor rises in water merely makes it weightless in a relative sense; in air the mixture's heaviness will immediately become apparent, in mercury it will immediately become light. So "weight" throughout this essay designates either the potential or active presence of heaviness, lightness, and all composite measures of them. Weightlessness designates the very absence of such presence, namely, the inability of the weightless to partake, naturally or by force, in up and down motion. Hence:

But the body that moves in a circle cannot possess weight or lightness; for it cannot move towards or from the centre either naturally or counter-naturally. For rectilinear locomotion does not belong to it naturally, since the locomotion of each single body is single, and so it will be the same as one of the bodies that move in this manner. Were it to move counter-naturally, then if movement downwards is counter-natural, movement upwards will be natural, but if movement upwards is counter-natural, that downwards will be natural; for we laid down that, with contraries, when one movement is counter-natural for a thing, the other, contrary movement is natural. Since the whole and the part (for instance, the entire earth and a small clod) move naturally to the same place, a first upshot is that this body possesses neither lightness nor weigh at all since otherwise it could move either towards or from the centre according to its own nature; next, that it cannot be moved spatially by being drawn upwards or downwards, for it cannot possibly be moved with another movement either naturally or counter-naturally, neither it nor any of its parts—the same argument for the whole as for the part." (*On the Heavens*, I.iii,269^{b}29-270^{a}12, Leggatt 1995)

Weightlessness in the above sense is the fundamental nature of Aristotle's ether that dominates the celestial region (from the sphere of the moon outward).

Weight is the fundamental nature of all terrestrial matter, below the sphere of the moon. Circles have no points that serve as unique starting and ending points; all straight line segments do. In contradistinction from the weightless ether that has no contrary in keeping with the character of a circle,[3] the realm of weight is defined between two extremes—the absolutely heavy and the absolutely light. Being absolute, they derive directly from the absolute size of the universe, which must, therefore, be finite to prevent infinite absolute heaviness and lightness.[4] Between them, all continuous transitions take place (and it should be understood that for Aristotle, the actively light is always potentially heavy, and vice versa, since both belong to the same family of straight motions). Earth is the absolutely heavy element. Fire is the absolutely light element. Aristotle considers that in order to ensure a continuous spectrum from the absolutely heavy (earth) to the absolutely light (fire) two and only two more elements are required, water and air. These reflect relative heaviness and lightness in the sense that regardless of size, a bubble of air always rises in water, and a drop of water always sinks in air, while fire always rises in air, and earth always sinks in water. Natural up and down motion, then, characterizes the terrestrial elements, and they are mutually transmutable by virtue of belonging to the same spectrum between absolute heaviness and lightness. The celestial element is characterized by circular motion. It cannot transmute into any of the terrestrial elements and vice versa. While terrestrial matter can be forced to move in concentric circles about the earth, celestial matter cannot be forced to move up or down. Terrestrial physics is therefore based on laws of motion that are distinct from the laws of motion that underlie astronomy, or celestial physics. Both belong to the realm of physics to the extent that they study matter in motion, and not to mathematics, which studies forms and magnitudes in abstraction from matter and without recourse to time.[5]

---

[3]This does not mean that Aristotle rejects the possibility of contrary motions in the ethereal realm. Indeed, he states explicitly (*Physics* VIII.viii.262^{a}9-12) that any circular motion has a contrary that will counteract and arrest it. This, however, is a contrariety that lacks a critical feature of linear motion, in which contrary motions go to contrary places, while in circular motion, all rotations and their contraries go from the same starting point to the same ending point.

[4]Since heaviness is actively expressed by the speed of a heavy object's natural motion, Aristotle argues that an infinite universe implies infinitely fast natural motions, which is absurd. More on this in Appendix A.

[5]I find this the only way to make consistent sense of Aristotle's reference to astronomy and optics as the more physical of the mathematical subjects: "...one must see how the mathematician differs from one who studies nature (for natural bodies too have surfaces and solids and lengths and points, about which the mathematician inquires), and whether astronomy is different from or part

A naturally falling stone, then, is not driven down by its heaviness. It does so by nature. Heaviness is not a force that pushes or pulls a stone downward, and in this sense it features as the formal cause of natural fall rather than its efficient cause. The efficient cause is primarily whatever transmuted a bulk of light matter into heavy matter, say, a bubble of elemental air into a clod of elemental earth. Once the transmutation is complete, the clod of earth will fall down by virtue of being earth, and not because some efficient cause called "heaviness" is forcibly driving it down. A secondary efficient cause of natural fall is that which removes a block that prevents the natural downward motion of a heavy object. In this secondary sense, heaviness is an efficient cause of natural fall to the extent that it overcomes any external blockage to the motion,[6] including in particular the ever-present impedance offered by the medium. Considering all of this, a heavy mobile's weight cannot present a load against itself since the heaviness is not a force that acts against the mobile in the first place. Any relationship between speed and weight (heaviness or lightness) in either natural motion or its direct contrary must enter the scheme in a way that is distinctly different from the way it enters forced motion as discussed in Levels 1, 2, and 3.

The picture changes with respect to the medium. Heaviness, according to Aristotle, is that with which a naturally moving mobile cleaves its way through a medium. The inescapable conclusion is that a mobile's heaviness acts as an external force against anything outside the mobile that stands in its way down. The medium, however, differs from other obstacles in its ability to absorb moving force from external original movers. Whether or not the moving force that a heavy object builds into the medium applies back to the original heavy mover remains to be seen (see further development ahead). Even under the assumption that the medium directly effects or contributes to natural downward motion it remains objectionable to suppose that the heaviness, which cannot reasonably act against itself as a load, does present a load against the medium's contribution to the motion that reflects the very nature of heaviness.

(Footnote 5 continued)

of the study of nature. For if it belongs to the one who studies nature to know what the sun and moon are, but none of the properties that belong to them in themselves, this would be absurd, both in other ways and because those concerned with nature obviously speak about the shape of the moon and sun and especially whether the earth and the cosmos are of spherical shape or not. The mathematician does busy himself about the things mentioned, but not insofar as each is a limit of a natural body, nor does he examine their properties insofar as they belong to them because they pertain to natural bodies." (*Physics*, II.2.193^{b}22-34, Sachs 1995). Astronomers and students of optics do not study the properties of circles and lines per se as the mathematicians do; rather, they are reasoning about planetary motions and rays of light *in terms* of circles and lines. Mathematicians, by contrast, may point to figures in the sand, but it is not sand figures that they reason about, but abstract geometrical ones independent of all material constraints. Once given, the curvature of a nose may be studied as an abstract mathematical figure, but its constraint as belonging to a nose is physically imposed and cannot be derived in the abstract (See *Physics*, I. iii.186^{b}21-22, and II.ii.194^{a}6).

[6]*Physics*, VIII.iv.255^{b}14-256^{a}3.

It appears, then, that the dynamics of terrestrial vertical motion differs from the dynamics of terrestrial horizontal motion in the way that weight (heaviness or lightness) figures into them. In vertical motion, whether reflecting natural weight only or a combination of weight and any additional vertically acting agency, weight is a source of motion, and figures as an additive component to vertically acting external agencies. As the next quote shows, Aristotle indicates this additive aspect of natural vertical motion. However, he says very little about terrestrial horizontal motion, and some minimal guidance with regard to horizontal motion must be interpolated to obtain a measure of theoretical completeness. Specifically, in horizontal motion, which is never natural and always forced, let the heaviness (or lightness) of a body figure in as a load, which is inversely proportional to speed. However, an individual who cannot lift a loaded wagon could still push it horizontally. Therefore, while load in horizontal motion bears direct proportionality to weight (heaviness or lightness), it differs from the property "weight" as inextricably associated with vertical motion. Vertical in this context means radially up or down, and horizontal means in a circle around the center of the earth. Both vertical and horizontal motion require contributions from the medium that provides the effective force and resistance to motion, and we have already seen that according to Aristotle, motion without a medium (namely, in a void) invariably leads to paradoxes. The following statement by Aristotle focuses on the role of the medium (air, in this case) as effective mover and how it combines with natural motion:

> All movement is either natural or enforced, and force accelerates natural motion (e.g. that of a stone downwards), and is the sole cause of unnatural. *In either case the air is employed as a kind of instrument of the action,*[7] since it is the nature of this element to be both light and heavy. In so far as it is light, it produces the upward movement, as the result of being pushed and receiving the impulse from the original force, and in so far as it is heavy the downward. *In either case the original force transmits the motion by, so to speak, impressing it on the air.* That is the reason why an object set in motion by compulsion continues in motion though the mover does not follow it up. Were it not for a body of the nature of air, there could be no such thing as enforced motion. *By the same action it assists the motion of anything moving naturally.* (*On the Heavens*, III.ii.301ᵇ20-30, Guthrie, 1939, my italics).

The scope of the above quotation goes clearly beyond the specific case of natural fall from rest, and we cannot escape the task of trying to find how natural and forced motions combine. Particularly instructive at this stage are two arguments by Aristotle, showing how the discussion of natural motion ties quantitatively with the principles of level 1 (that is, *Physics* VII.v):

> That some bodies must owe their impulse (ῥοπὴν) to weight or lightness can be shown as follows. (i) We agree that they move of necessity. But if that which moves has no natural impulse it cannot move either towards or away from the centre. Suppose a body A to be weightless, and another body B to have weight, and let the weightless body move a distance CD and the body B move in an equal time CE. (The heavy body will move farther.) Now if the heavy body be divided in the proportion in which CE stands to CD (and it can quite well

---

[7] πρὸς ἀμφότερα δὲ ὥσπερ ὀργάνῳ χρῆται τῷ ἀέρι

bear such a relationship to one of its parts), then if the whole traverses the whole distance
CE, the part must traverse CD in an equal time. Thus that which has weight will traverse the
same distance as that which has none, and this is impossible. The same argument applies to
lightness (*On the Heavens*, III.ii.301^{a}24-301^{b}1, Guthrie 1939).

The second argument considers the possibility that the motion of a weightless
body is forced, and hence obeys the principles of motion under force. If so—
Aristotle puts down as claim to be proved—then at any finite time the weightless
body will be moved to infinity: "Moreover, if there is to be a moving body which is
neither light nor heavy, its motion must be enforced, and it must perform this
enforced motion to infinity" (Guthrie). The proof goes as follows:

That which moves it is a force (δύναμίς τις), and the smaller, lighter body will be moved
farther *by the same force*. Now supposed A, the weightless body, is moved the distance CE,
and B, the heavy body, the distance CD in an equal time. If the heavy body be divided in
the proportion in which CE stands to CD, the part cut off from it will as a result be moved
CE in an equal time, since the whole was moved CD. For *as the greater body is to the less,
so will be the speed of the lesser body to that of the greater*. Thus a weightless body and a
heavy body will be moved an equal distance in the same time, and this is impossible.
Seeing, therefore that the weightless body will be move a greater distance than any other
given body, it must travel to infinity. The necessity for every body to have a definite weight
or lightness is now clear. (*On the Heavens*, III.ii.301^{b}5-18, Guthrie, 1939, my italics).

The first thing to note about these two arguments is that neither constitutes a demonstration of inherent paradox in the concept of weightlessness falling naturally or by
force. What they do show is the incompatibility of weightlessness falling naturally or
moved by force with (1) the proportion relating heaviness to natural speed of fall;
(2) with the proportion relating heaviness to forced speed; and (3) with the finite size of
the universe. The two proportions and the finiteness of the universe are here taken for
granted. The reader is assumed to know and accept them, and then it is quickly shown
that under these fundamental restrictions, certain things are disallowed.

The first argument in particular betrays Aristotle's fascination with mathematical
physics: "We agree that they move of necessity. But if that which moves has no
natural impulse it cannot move either towards or away from the centre." This is so
by basic postulates that are already assumed for this discussion, and nothing further
needs to be added by way of proof. The redundancy of the mathematical proof that
Aristotle supplies underscores the importance that he attached to elaborating the
argument in mathematical terms.

The two arguments, which follow one another in quick succession, are the closest
Aristotle comes to teaching how to combine the principles of forced and natural
motion. The demonstrations follow his usual habit of representing the relevant
magnitudes by the lengths of line segments (not supplied by diagram with the text in
this case, but they are clearly intended and easy to draw). The importance of these
examples to our discussion warrants a closer look. Consider first the generalized
principle italicized in the second case: When two objects of different weights (say
heaviness, to keep things unambiguous) are moved by the same force, then their
weights and speeds are inversely proportional, that is $W_1:W_2::V_2:V_1$. Note that while
the example compares distances covered in equal times, the general principle

(italicized) is formulated in terms of speeds. This justifies our use of speed in levels (1) and (2). The argument is clearly intended generally, and so we may as well take the proportion to hold under all conditions, for variable as well as uniform speeds.

Keeping this in mind, consider the first case. Here we have a weightless body A falling naturally at speed V_A, and a heavy body B falling naturally at speed V_B. We represent the heaviness of B and the speeds V_A, V_B, by the respective parallel line segments FH, EJ, EI (Fig. 2.1):

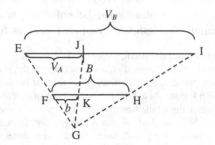

Fig. 2.1

Extend the lines through EF and IH until they meet at G. Draw JG, cutting FH at K. Then clearly, where EI = V_B, EJ = V_A, FH = B, and FK = b:

$$\frac{V_B}{B} = \frac{EG}{FG} = \frac{V_A}{b} \Rightarrow b : B :: V_A : V_B$$

This demonstrates Aristotle's contention that a part b of B can always be marked such that its ratio to B is as the ratio of the weightless body's assumed finite speed V_A is to B's speed, V_B. The motivation to construct this proportion in the first place is the postulate that in natural fall, the weights relate directly as the speeds, and then follows the inevitable conclusion that no finite speed, however small, can be assigned to the weightless body, which cannot, therefore, fall at all.

Turning to the second example, draw OE = V_B and OF = V_A. Extend FO to H and let OH be the weight of B (Fig. 2.2). Draw from H a line parallel to EF, and mark by G its intersection with the extension of EO. Let the length of OG be b, and clearly, by the constructed similarity of triangles OEF and OGH:

$$V_A:V_B::B:b.$$

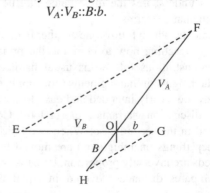

Fig. 2.2

Once again, a simple construction vindicates Aristotle's claim that given speed V_A forced upon a weightless body A, and speed V_B generated in a heavy body B by the same force, a body of heaviness b may always be found such that the two finite weights are inversely as the two given speeds. To the extent that such is the proportion of heavy weights to speeds under the effect of the same force, it turns out that the speed of the weightless body must be increased indefinitely, or else there will always be a heavy object moved at the same speed by the given force. But an infinite speed means a universe of infinite extension, contradicting the finite size of the universe, which is here taken to be beyond dispute.

Note that the weights and the speeds need not be drawn using the same units of length. Speed lines may be measured in millimeters, weight lines in inches, and the equality of ratios will remain unaffected as long as each quality has its standard of length. That is to say, once the lengths OE and OF have been drawn proportionate to the given values of V_B and V_A respectively, OH representing B may be drawn to any arbitrary length, and the procedure ensures that its ratio to OG representing b will remain invariable.

The difference between forced and natural motion should be quite clear. In natural fall, heavy weights and speeds relate *directly*: $W_1:W_2::V_{n1}:V_{n2}$. The speeds of the same W_1 and W_2 moved by some given force are *inversely* as the weights: $W_2:W_1::V_{f1}:V_{f2}$. (V_n and V_f stand respectively for natural speed and forced speed).

Proper combination of the principles of forced motion and the principles of natural motion must specify how force, mobile, and motion relate along the line of natural motion and along a line transverse to the line of natural motion. With respect to vertical motion, two versions can be constructed for the interaction between a body's heaviness and the medium as effective mover. While different in their mode of explaining the vertical motion associated with heaviness (or lightness) combined with some external moving agency, the resulting motion is identical in both cases.[8] In version (a), the natural heaviness impregnates the medium with effective moving force that causes active movement. Here, natural heaviness competes with any external upward force in the medium, and to the external mover this competition manifests itself either as resistance of a heavy object to forced upward motion (or as resistance to downward motion in excess of the motion due to its inherent heaviness). In version (b), heaviness does not impregnate the medium with an effective force that moves the body downward, and its active nature manifests itself directly as downward motion. In other words, downward motion is not the result of some dynamical interaction between heaviness and the medium as effective mover; rather, downward motion is fundamentally the active expression of heaviness. From this stems the resistance of a heavy object to an external lifting agency as it impregnates the medium with effective upward moving force (or to an external agency that attempts to impregnate the medium with effective force to

---

[8]See Appendix D for a mathematical demonstration that the two version lead to identical descriptions of motion.

speed the heavy body beyond its natural downward speed). The following points further clarify the main features of the two versions:

1.

a. Heaviness unopposed by the upward directed action of some external agency operates on the medium in a downward direction as an external agency, stressing to cleave it while infusing it with effective moving power, in accordance with level 3 above. Heaviness, then, competes *in the medium* as a downward agency against any opposed external agency. The net intensity of the source that builds effective moving force into the medium is proportional to the difference between the internal heaviness and the overall intensity of any external opposing agency. Therefore, only an external agency greater than the heaviness of a mobile will have an excess over it that can infuse upward effective force into the medium, moving the mobile upwards against its weight. The excess over natural heaviness, then, acts as the power of an original mover building a net upward moving force in the medium. Constant upward speed will obtain once the rate of force decay in the medium matches the infusion of effective force into the medium. Once contact between the external agency and the mobile is broken, the built-up force in the medium will gradually decay with no counteracting infusion, sustaining a decelerating upward motion until it dissipates altogether. But the process will not cease here, because the heaviness constantly strives to build downward effective force into the medium. Therefore, downward motion will take over with accelerating rate while downward driving force builds up in the medium under the effect of heaviness. Terminal speed is reached when the dissipation of force in the medium matches the rate of force infusion into it by the downward force of heaviness. An external agency acting downward, then, builds into the medium downward effective force beyond what the mobile's natural heaviness does. Once disconnected from the mobile, the downward motion will decelerate back to the speed associated with the effective force that the heaviness builds into the medium by itself. At all stages, however, whether the motion is purely natural or a combination of natural and unnatural components, the effective force in the medium is operating, and in this sense all motion, natural or not, has a forced aspect.

b. Heaviness is a principle of motion internal to the body. To external bodies, including the medium, it appears as an external moving agency. To the heavy body itself, the medium manifest itself by its resistance to being cleaved, and by the additional resistance reflected against a heavy body as it impregnates the medium with effective force to move external objects *other than* itself. In this case, the effective force that an external upward source infuses into the medium acts directly against the inherent principle of motion that the natural heaviness of the mobile represents. Once again, even when supporting a motionless object, an external force must exert itself to keep inflow of upward effective force into the medium so as to exactly counteract the natural heaviness of the object. In this rendition, a heavy body moves up only once the surrounding medium contains an effective force greater than the body's heaviness. Upward motion will first accelerate as the excess over weight in the medium builds up, and continue at constant speed once decay matches input at this higher level of effective force. Upon removal of the external source, the upward effective force in the medium will decay, as always. When its continuously dropping level exactly balances the natural heaviness of the mobile, the mobile will pass through an instant of rest, but downward motion will follow at accelerating rate as the effective force drops further and further below the natural heaviness. Terminal speed settles in once all the effective force is exhausted, with the magnitude of the speed directly proportional to the heaviness, and inversely to the resistance of the medium to being cleaved. Note again that in this version heaviness does not infuse the medium with effective force to move itself. Therefore, under this understanding of the dynamics, natural motion is unforced—the medium is cleaved by the heaviness and the resulting speed reflects only the heaviness mitigated by the resistance of the medium to being cleaved and the additional difficulty of loading it with effective force to move other objects. The motion of a heavy body either hurled or released from rest will be exactly the same in both versions. But physically, while in version (a) the medium builds up downward force to account for the acceleration, here it is the difference between the decaying upward force in the medium and heaviness that accounts for the downward acceleration. Under this interpretation, when Aristotle speaks of force aiding natural motion, he should be understood as referring to downward external agencies building extra effective downward force into the medium to directly aid the natural heaviness. This alternative is worth keeping in mind because it corresponds exactly to Hipparchus's account of prolonged motion, with the possibility of transferring the medium's active role into the body, and leaving the medium only with a purely resistive effect.[9]

2. So far, it seems plainly clear that the transition from upward to downward motion is perfectly smooth, passing through rest at a single instant of time, so that the body never rests for any duration, no matter how small. However, this aspect will require an important modification following the discussion concerning the threshold of motion in the next section.
3. In a direction transverse to the line of natural motion, load is proportional to the heaviness of the body moved, so that speed is directly proportional to the external mover's strength, and inversely proportional to the load. This does not

---

[9]For a more detailed comparison of Aristotle's active medium to the idea of in-body impetus, and whether Hipparchus really originated the idea, see ahead, "Hipparchus on the Theory of Prolonged Motion."

imply that a body's load against horizontal motion is quantitatively equal to the body's heaviness. The physical distinction between the behavior of a heavy weight in vertical motion and its related load in horizontal motion must not be forgotten: even an extraordinarily powerful human cannot lift up vertically a packed cart weighing half a ton, but it takes no more than average human strength to push it along a horizontal floor. However, since all relations without exception take the form of proportions, the ratio of two heavy loads is invariably the same as the ratio of their respective weights, and it becomes a matter of indifference to use weight ratios only.

When a hurled object moves in a direction other than radially up or down, the radial and horizontal aspects of its motion must be analyzed separately according to the principles outlined above, and then combined into a full account of the result. There is no room for plausible doubt with regard to Aristotle's ability to combine directed magnitudes. In Eudoxan homocentrics, which Aristotle describes in *Metaphysics* XII.viii, velocities are directionally superimposed as a matter of course. In *Physics* VIII.viii.262^{a}13 Aristotle indicates that motion up is not the opposite of motion sideways, and they cannot cancel each other. *Meteorology* I. ii.339^{a}20-32 describes how the circular motion of the ether confers a forced circular (horizontal) component on the natural vertical motion of the terrestrial elements, providing the source of the variable motions observed in the terrestrial region. The *Mechanical Problems*, presumably written by Aristotle's students, clearly teaches the parallelogram method of combining directed magnitudes, and also argues very cogently that when a constant horizontal speed is combined with a variable vertical speed, the result is motion along a plane curve. Given all of this, it is implausible to assume that Aristotle was incapable of combining all manners of vertical and horizontal motions.

Figure 2.3 is a qualitative graphical representation of the stages of upward flight. At t_0, an upward moving agent that has been holding an object at rest against its heaviness suddenly increases power beyond the object's weight. At t_1, it breaks contact with the object, which behaves as a thrown object from this point on. The force curve shows the resulting development of effective moving force in the medium for version (a), and the same curve shows the difference between the body's heaviness and the changing effective force in the medium for version (b). The associated speed and position curves illustrate the resulting motion. When marked above the time-axis, the magnitudes of the effective force and the speed represent upward tendencies; when marked below the time-axis, they represent downward tendencies. Position above the time-axis represents position above the throwing platform from which motion began; position below the time-axis represents position below the throwing platform.

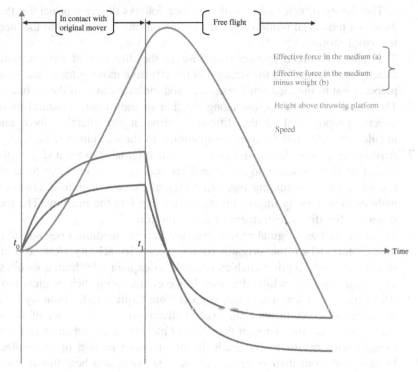

Fig. 2.3

Summary: The Main Features of Aristotle's Theory of Terrestrial Locomotion

We now have the basic framework of Aristotle's theory of matter in motion. The main features, in brief outline, are as follows ("motion" throughout this summary refers exclusively to spatial displacement):

1. All natural motions are unforced. All unnatural motions are forced.
2. For all terrestrial mobiles, natural motion is either radially toward the center of the universe (heavy mobiles) or radially away from the center of the universe (light mobiles).
3. Accordingly, all terrestrial mobiles are classified along a common scale of weight, from the absolutely heavy earth, to the absolutely light fire.
4. Weightlessness is the inability to partake in either the heavy or the light. Such is the nature of the celestial ether, which cannot, accordingly, change its distance from the center of the universe, and can only move on spherical surfaces around

it. The theory of celestial motion therefore follows different principles than the theory of terrestrial motion. The rest of this outline is restricted to the theory of terrestrial motion.

5. The speed of forced motion transverse to the direction of natural motion is directly proportional to the strength of the effective moving force, and inversely proportional to the medium's resistance and the magnitude of the mobile's load.

6. The speed of forced motion along the line of the mobile's natural motion is directly proportional to the difference between the effective force and the mobile's weight, and inversely proportional to the medium's resistance.

7. Aristotle makes a clear distinction between original movers and the effective force that they generate to move mobiles actively. The effective force always resides in the surrounding medium. Original movers, therefore, always move mobiles indirectly by impressing effective force into the medium. The medium is always the direct instrument of active motion.

8. The force that an original mover impressed on the medium does not disappear immediately when the original mover ceases its activity. Rather it decays slowly. This assumption enables Aristotle to explain why hurled mobiles continue to move for a while after they broke contact with their original movers.

9. All of the above features obtain support from explicit indications by Aristotle. Aristotle does not provide an explicit discussion of the onset of motion to complement his discussion of the decay of motion. In continuation of his line of thought with regard to the gradually decelerating motion of projectiles after breaking off from their original movers, it is postulated here that it also takes time for original movers to build impressed force into the medium. This complements Aristotle's account by matching the gradual deceleration of forced motion with a complementary gradual acceleration under the influence of an original mover.

Certain refinements are still necessary, beginning with the most important qualification of the theory so far—the threshold of motion.

Chapter 3
Some Refinements of the Basic Theory

The Threshold of Motion

As we already saw, in *Physics* VII.5 Aristotle explains that if effective mover A moves mobile B over distance C in time D, then half of A (call it E) moves half of B (call it Z) over distance C in time D. The same E (½A) will move two Z loads (each being ½B), over half the distance C in time D. This is because E splits its time between the two loads, so each of them gets only half the time. However:

> But if A will move B over the whole distance C in time D, half A (E) will not be able to move B, in time D or in any fraction of it, over a part of C bearing the same proportion to the whole of C that E bears to A. Because it may well happen that E cannot move B at all; for it does not follow that if the whole force could move it so far, half the force could movie it either any particular distance or in any time whatever; for if it were so then a single man could haul the ship through a distance whose ratio to the whole distance is equal to the ratio of his individual force to the whole force of the gang (*Physics* VII.v.250ª12-19, Wicksteed and Cornford 1929).[1]

Motion, then, has a threshold: if the mover's force falls short of this threshold, then despite being applied to the mobile, it will generate no motion. If the mobile can be divided into n equal loads, and if the mover can move one of these loads over distance G in time D, then all n loads can be moved one at a time to a distance G/n in time D, or over distance G in time nD.[2]

---

[1]In a footnote, Cornford adds "[*Literally*, 'otherwise one man could move the ship, since both the force of the haulers and the distance of which all of them together make it move are divisible into the (same) number (of parts as there are men).']".

[2]This does not take into account the time taken for the original mover to go back to the starting point after transporting each of the partial loads. However, since the relation here pertains to the effective moving force in actual operation on the loads and not to any particular original mover, this omission is justified.

© The Author(s) 2015
I. Yavetz, *Bodies and Media*, SpringerBriefs in History of Science
and Technology, DOI 10.1007/978-3-319-21263-0_3

This is quite simple and straight forward, and yet the qualification is sometimes overlooked, creating unnecessary difficulties.[3] It also requires a modification of the theory of motion developed in the previous sections. Figure 3.1 is a qualitative graphic representation of an upward throw incorporating the correction required by the threshold of motion, reflecting versions (a) and (b) of the previous section. Up to t_0, a moving agent has been infusing the surrounding medium with effective moving force to exactly counteract the mobile's heaviness. At t_0 the moving agent suddenly increases its force beyond the mobile's heaviness. At t_1 contact between the moving agent and the object breaks, and the effective moving force decays. The force curve shows the effective moving force in the medium (a); or the difference between the growing effective moving force in the medium and the mobile's intrinsic downward tendency, that is, its heaviness (b). The net effective force is directed upwards when its magnitude is marked above the time axis, and downwards when marked below the time axis. Two horizontal lines mark the magnitude of the threshold of motion, E_{th}, and the vertical lines that cross the time axis at t_a, t_b, t_c, mark the times when the varying net effective force crosses the threshold. The speed and position traces show the motion that results from the combined effects of the force and the threshold of motion. Specifically, they feature the period of rest between upward and downward motion. This graph should be compared with the one concluding the previous section on the theory of vertical motion (Fig. 2.3).

---

[3]For example, in *On the Heavens* II.vii, Aristotle explains that the celestial bodies (stars, planets, sun and moon) are made of ether, and hence do not burn and do not shine. They generate heat and light by chafing diffusions of air in the ethereal region through which they move (for a well-argued claim that such diffusions do not contradict any element in Aristotle's cosmology, see J. Thorp 1982, pp. 104–123). Starlight and sunlight, then, are byproducts of friction. In *On the Heavens* II. ix, however, Aristotle rejects the Pythagorean suggestion that the stars produce harmonious music as they move. This cannot be so, Aristotle says, because the stars are carried in their ethereal spheres, and hence do not move through a stationary medium. Had they been moving through air, he says, the noise would indeed be deafening, given the enormous speeds involved. Guthrie, for one, cannot see a way out of the overt contradiction between Book II.vii on the friction-generated light of the stars, and book II.ix that seems to reject any such friction (*On the Heavens*, Guthrie, 1939, note *a*, p. 196). However, the difficulty dissolves under the assumption that the threshold for light and heat generation is lower than the threshold for sound generation: the motion may generate sufficient friction to ignite the terrestrial diffusions encountered by the planet, but not sufficient to produce sound, possibly because the diffusions are so rare. That sound-generation will not occur below a certain threshold is explicitly stated by Aristotle in *Physics* VII.v: "And in this lies the fallacy of Zeno's contention that every grain of millet must make a sound as it falls (if the whole measure is to do so). For it may well be that in no period of time could the one grain move that air that the whole bushel moves." (*Physics*, VII.5. 250ª20-23, Wicksteed and Cornford 1929).

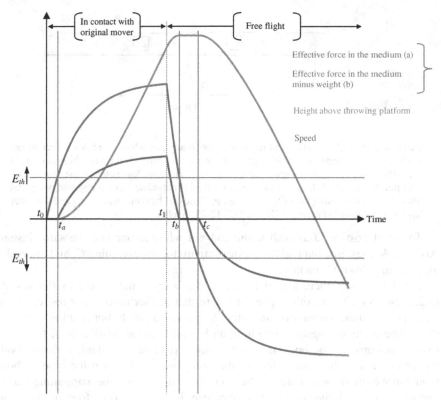

Fig. 3.1

Physics VIII.iii.253^{b}14-30 and possibly also the concluding part of Book VII.v add a time threshold to the force threshold. In other words, a force that causes some change when applied for some time, may not cause any change if applied half the time. Detailed analysis of this additional aspect creates too much of a digressive break from the main argument in this essay. For further discussion, see Appendix B.

Segments of Time, Instants of Time, and the Concept of Speed

Now the point belong in common to what is before and what is after, and is the same and one in number, but it is not the same in meaning (since it is the end of one, the beginning of the other); but for the thing it always belongs to the later attribute. Call the time ACB and the thing D. This is white during time A, but not white during time B (Fig. 3.2);

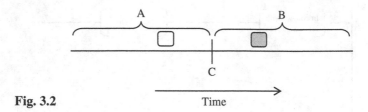

Fig. 3.2   Time

therefore at time C it is white and not white. For in any part whatever of A it is true to say that it is white, if during this whole time it was white and during B it is not white; but C is in both. Therefore one must not grant the 'during all…' but must leave out the last now, which is C; but this already belongs to the later time. And if not-white was coming into being and white dying away during all of time A, then at C the one has come into being and the other has been destroyed (*Physics*, VIII.viii. 263^{b}12-23, Sachs).

To avoid posting D as both white and not white at one and the same instant, Aristotle defines time interval A as extending up to, but excluding C, and interval B as extending from C, inclusive.

Is it fair to say, then, that at instant C the body is not white? In *Physics* VI. iii.234^{a}24-234^{b}10, Aristotle argues at length that neither motion nor rest can *be* at an instant of time. Instants of time may be passed through, but nothing can *be* in them. The argument against motion at an instant uses the possibility of faster and slower motions. If one body moves to some extent in an instant, a slower body would move less than that extent in the same instant. But then the quicker body would move the same distance as the slower one in a lesser time, suggesting that an instant of time is divisible, which is, of course, impossible. Therefore, motion in an instant is not possible. Furthermore, Aristotle continues, since rest is attributable only to things that have the ability to move, the argument still applies, and instantaneous rest is equally impossible as instantaneous motion. At an instant, things cannot be either in motion, or at rest. An instant is a cut in time; it may be passed through, but nothing at all can ever *be* in it.

In the transition illustrated by the diagram above, C marks the beginning of the period of D's non-whiteness. But no first period of non-whiteness can ever be singled out. Aristotle shows this easily, under the assumption that time is always divisible. (In this passage, when Aristotle speaks of "time," he clearly has in mind a period of time, bounded between two "cuts," or instants of time) (Fig. 3.3):

Fig. 3.3

And further, if we say something has been moved in the whole time XR, or in any time in general, by coming to the last now of it (since this is the boundary, and what is between nows is time), then also in the same way it should be said to have moved in the other times. And the division {K} is the extremity of its half. So it will have been moved also in the half time and in general in any of the parts whatever, since by means of the cut, every time is always bounded by the now there with it. Then if every time is divisible, and what is between the nows is a time, every changing thing will have undergone an infinity of changes (*Physics*, VI.vi.237ª3-11, Sachs 1995).

In other words, if we perceive something as moving, it must have already been in motion before. We cannot, therefore, isolate an initial period in which motion first began, because any such period may be subdivided, yielding an earlier initial period in which motion already existed. We can only identify the first instant from which in any period, no matter how short, the body is already moving.

While Aristotle does not appear to have a separate, systematic discussion of speed, he often speaks of motions being faster or slower. One motion is faster than another if it covers the same extent at a lesser time than a slower motion; if in the same time it covers a larger extent than the slower; and if in a lesser time it can still cover a larger extent than a slower motion, but not so much larger than the slower motion as it does at the same time (*Physics*, VI.ii.232ª23-27). Such distinctions always relate to periods of time, and Aristotle never reduces them to a limiting statement concerning instantaneous speed. Indeed, in Aristotle's way of thinking it is difficult to see how speed, being an attribute of motion, can exist at an instant when motion itself cannot.

Having said all that, Aristotle clearly distinguishes between uniform and variable motions. In terms of path, only straight and circular motions are uniform, he says, but uniformity and variability can also be identified with respect to speed:

> But sometimes the non uniformity is not in that which moves, nor in the when, nor in the direction, but in the way it moves. For it is sometimes distinguished in quickness and slowness, and that of which the speed is the same is uniform, that of which it is not, non-uniform. Hence speed and slowness are not kinds of motion, nor specific differences, because they go along with every specific difference of kind (*Physics*, V.iv.228ᵇ26-30, Sachs 1995).

Combine this with the endless divisibility of time, and Aristotle's view may be used to examine non-uniform motion at any desired finesse.[4] He does not produce any mathematical study of distances covered by types of non-uniform motion, but the understanding of speed variations above certainly opens the way to mathematical studies of motions at non-uniform speeds. In particular, the absence of the concept of instantaneous speed hardly poses a conceptually insurmountable obstacle, though the absence of something like the integral and differential calculus certainly poses very severe practical limitations.

---

[4]Time (also space) for Aristotle, does not have an existence independent of matter. Indeed, it is material motion and its variability that renders necessary the continuity of time: "And since every motion is in time, and in every time it is possible for a thing to be moved, and every moved thing admits of being moved both faster and more slowly, then in every time the faster will also be able to move more slowly. These things being so, it is also necessary that time be continuous. I call continuous that which is always divisible into divisible parts, for once this is set down about the continuous, time must be continuous." (*Physics*, VI.ii.232ᵇ21-26, Sachs).

Formal and Dynamical Objections to Motion of Unlimited Extent

Nor yet, fifthly can it {motion} either accelerate or slow down for ever, for if so the movement would be infinite and undetermined, whereas we believe that every movement is from one fixed point to another and is determinate (*On the Heavens*, II.vi. 288[b]29-30, Guthrie 1939).

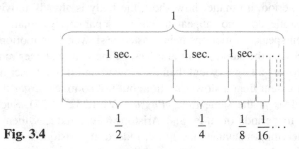

Fig. 3.4

In a finite universe like Aristotle's the objection to endlessly accelerating motion is clear: it leads to infinite speeds that necessarily imply an infinite supply of space. But what of a mobile that covers half a meter in the first second, a quarter of a meter in the second, an eighth in the third, etc.? (Fig. 3.4)

Such a motion has a very definite distance limit of 1 meter—no more, and no less, so the objection to infinite distances that undermines endless acceleration does not apply. Aristotle knows this very well, but still rejects the possibility of such endless deceleration. While his cosmos is limited in space, it is not limited in time; it was never created, and will never vanish into nothingness; it always was, and always will be. As such, it provides an infinite supply of time. Why, then, object to the possibility of the decaying motion above?

One of Aristotle's arguments against the possibility of an infinite rotating body reveals the source of the present objection. Imagine a semi-infinite line, rotating around a point C close to its definite end A (Fig. 3.5). Below point A mark the infinite line BB, and consider the rotation of the semi-infinite line from the moment it is perpendicular to BB:

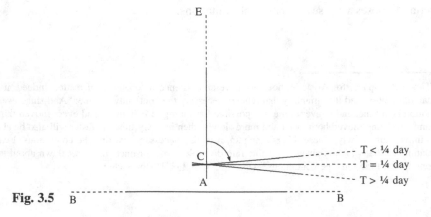

Fig. 3.5

Assume, with Aristotle, that the semi-infinite line AE rotates about C once every sidereal day, like the stellar sphere. Clearly, at any time, no matter how brief before one quarter of a sidereal day, the two lines do not yet cross, and at any time after a quarter of a sidereal day, they already cross one another somewhere at a finite distance to the right of C. Crossing must, therefore, begin exactly one quarter of a sidereal day after perpendicularity.[5] This settles the "when" of first crossing. But we cannot specify where, in space, the first crossing occurs. In the diagram, consider the lines a very brief finite time past one quarter of a sidereal day. They are already crossed, and the distance to the point of crossing, no matter how far from C, is finite. Subtracting that distance from the infinite length of AE leaves an infinite length. This means that the point in space of first crossing is still infinitely far away from the well-defined crossing point that occurs a little after one quarter of a sidereal day. So, while we can approach the *time* of first crossing as closely as we like, this does not get us nearer *in space* to the point of first crossing, which remains indefinitely far away. Now consider Aristotle's requirement as he begins this discussion of a semi-infinite rotating line:

> Further, if you subtract a limited time from a limited time, the remainder must be limited and have a beginning. If, however, the time of the journey has a beginning, there is also a beginning of the movement, and so of the magnitude covered as well (*On the Heavens*, I. v.272^{a}7-8, Leggatt 1995).

Aristotle requires that for a motion to be definite, it must be definite in both time and space. But the semi-infinite line's rotation is determinate in time only, not in space, and "...it is therefore not possible for the unlimited to turn about in a circle." (*On the Heavens*, I.v.272^{a}20, Leggatt 1995) By implication, therefore, since the visible sky clearly does rotate above us, it cannot possibly be infinitely distant.

It is now easy to see that the decelerating motion in the previous case represents the flip side of the one above. The motion of the infinite rotating body is determined in time, but not in space; the motion of a body whose speed decays in a geometrical series is well determined in space, but not in time. Aristotle requires that if a motion is well determined in time, it must also be well determined in space, and vise-versa. Therefore, since geometrically decelerating motion is spatially determined, it cannot persist over an indefinite time span, and must be cut off at some point within a finite time of the start.

This argument against endlessly decaying motion follows from investigating purely formal aspects of motion, namely its path and limiting points in time and space. The efficient aspect of Aristotle's dynamics of material locomotion, (which combines material, efficient, and formal aspects), nicely complements it. Here the necessary cutoff of any decelerating motion follows from the requirement that for

---

[5]Recall how Aristotle discussed the object that is white prior to a given instant and not-white afterwards: up to, but not including that instant, the body is not done turning non-white. From that instant on, inclusive, it already is non-white. In the case of the rotating semi-infinite line AE, up to, but not including one quarter of a day after perpendicularity, it is not yet crossing BB; it only approaches crossing. From one quarter of a day, inclusive, it already crosses BB.

any given load to be moved, a threshold of minimal force needs to be exceeded, and force represents the efficient aspect of the dynamical account. To the extent that the speed diminishes, so must the effective moving force, according to the basic relations of level (1). At some point, the decaying force will fall under the threshold of motion, and the mobile will come to a full stop.[6] The dynamical theory of locomotion, then, adds detail in revealing the direct efficient cause of a cutoff, previously made necessary by a formal aspect of well-determined motion as such. This harmony between dynamical considerations on the one hand, and formal considerations irrespective of efficient specifics on the other, is philosophically pleasing when successfully implemented. However, as we shall see in Aristotle's observations with regard to reciprocating motion, the need to match efficient and formal considerations can also become a source of endless frustrations.

Reversal of Motion and Collisions

Creation out of nothing and destruction into nothing are equally impossible to Aristotle. The natural motion of earth is the physical contrary of the natural motion of fire. Therefore, *actively existing* earth as an operating, active being cannot turn into *actively existing* fire, and *actively existing* fire cannot turn into *actively existing* earth—both transitions involve the absolute annihilation of one thing, and the creation out of nothing of another, because a thing cannot become its own contrary. And yet, Aristotle constantly maintains that such transitions regularly occur between the elements. His way out of positing creation out of nothing and extinction into nothing is in the distinction he proposes between the *active* and the *potential*. Actively existing earth is at one and the same time potential fire, and vise-versa. The transition of earth into fire is not one of active earth becoming active fire, but rather of potential fire becoming active fire, while active earth turns into potential earth. So Earth, as such, is never created and never annihilated—it only transits between its active and potential phases. Had earth not possessed within

---

[6]A measure of ambiguity still remains, since the process of force decay irrespective of any locomotion it might engender, is itself "motion" in Aristotle's general sense of any timed change. If the formal argument given in *On the Heavens* holds generally, then the force decay itself should be cut off and not continue indefinitely, and we now have a competition of thresholds. However, motion will stop when the first one is reached, and nowhere does Aristotle's text help us figure out which will be the actual cause of the secession of motion. Of course, while treating the formal aspect of force decay as a type of change, remember that it must also have its own efficient aspect, and now we end up caught in a seemingly endless regression. For treating locomotion and its efficient aspect, we need only recall the following. (1) In the basic discussion of the threshold of motion in *Physics* Vii.v, a single person applies force to a ship, but fails to generate motion. (2) In Aristotle's discussion of prolonged projectile motion in *Physics* VIII.x, once the force in the medium falls below the ability to generate more moving force and can generate motion only, the whole progression stops, and no further motion ensues. In studies of locomotion, therefore, we may ignore higher orders of thresholds.

itself the potentiality of fire, the transition of an active clod of earth into an active bulk of fire would never take place.[7]

Factor in time, and consider a hypothetical sudden transition of an earthen object into fire. Up to, but excluding instant A, the object was active pure earth. From instant A, including instant A, the object is active pure fire. This, however, does not reveal the complete picture as Aristotle proposes to see it. Through each instant of time and during each segment of time up to but not including A, while the object was purely active earth it was also purely potential fire. From instant A inclusive, the object is pure active fire and also pure potential earth. There is not a single instant of time, let alone a segment of time, in which the object is both active earth and active fire. At no time does Aristotle's view encounter the problem of positing a thing that is at one and the same time what it is not, because at any given instant the opposites occupy the distinctly different levels of the potential and the active. Gradual transitions work out in the same manner, allowing for different parts of a continuously divisible bulk to make the transition as time goes by.

Now consider that the physical essence of the element earth is downward motion, and the physical essence of fire is upward motion. To the extent that earth is always also potentially fire, the active physical essence of earth— downward motion, is potentially also the physical essence of fire—upward motion. As a rising bulk of fire gradually transitions into earth, its upward motion will slow down— reflecting the relative abundance of active fire and active earth in the mixed volume. At some stage, the locomotion will reverse, and if the process of change from fire to earth is continuous, the reversal will pass through a single instant (of no duration) of rest. At no instant do we have active fire and active earth at the same time and at the exact same points in space. And hence, at no time is the motion both actively upward and actively downward, but only the active end result of the competition between them. Rest, in this process, characterizes a single instant of time, and as Aristotle says, things can only pass through an instant of time, but never *be* in it. To be in time always means to be in a period of time—a time interval between two distinct instants, no matter how close.

All of this, however, does not take into account the threshold of motion. To the extent that the transition from fire to earth is continuous, the moving capacity associated with the difference between their respective quantities will have to fall below the threshold of motion generation. Ignoring for the moment the air as the effective mover, this means that during some brief period of time the transition of fire to earth continues below the threshold of motion, and for the duration of this period, the transforming body must freeze in space. The effect of the air only shifts the time and location of this rest period by its prolonging effect, but in time, the

---

[7]Strictly speaking, the properties of all matter, including the elements, are carried by the fundamental substrate of material properties—the *hulé*. The *hulé* that carries the properties of a clod of earth actually, also carries those of fire potentially. Since fire and earth are opposites, the *hulé* cannot carry both simultaneously in actual state, and the transition of potential fire into actual fire in the same bulk must be simultaneously accompanied by the transition of actual earth into potential earth.

effective moving force of the air will have to reflect the transition in the original moving agency. To the extent, then, that every terrestrial motion without exception has a motional threshold, the transforming body in this example will have to come to a full stop, and maintain a state of rest for a brief period of time before beginning its active descent. The downward motion will start only once the excess of earth over fire loads the surrounding air with effective moving force that overcomes the threshold of motion.

The case of a clod of earth hurled upward differs from the previous case because here the moving body retains its elemental constitution throughout the motion.[8] This requires the theory of projectile motion, as summarized in level (3). As the motion up continues past contact with the original upward hurling mover, the consecutive layers of air around the mobile lose effective force continuously. Once the force falls below the motional threshold, motion will cease, and rest ensues while the upward force of the air continues to diminish, equalizing and then falling below the natural downward tendency of the body, until the difference exceeds the downward threshold of motion. The strictly continuous process of force decay in the medium is never broken, so that the clod of earth passes through the experience of no net tendency just as it passes through a single instant of time—it passes through it without occupying time there. However, the up and down locomotion associated with the continuous force change is discontinuous, with the up and down phases separated by a period of rest owing to the constraint imposed by the threshold of motion.

Having understood all that, we are in position to review one of Aristotle's most frustrating requirements, which drew the disapproving fire of many critics through the ages:

> But it is most clear of all that motion in a straight line cannot be continuous because what turns back must stop, and not only on a straight line but even if it is moved on a circle. For it is not the same thing to be moved in a circle and on a circle; for it is possible sometimes to go on being moved, sometimes to turn back again upon coming to the same place from which it set out. The belief that it must stop rests not only on the senses but also on reason. [...] For if H were carried to D and, turning back again, were carried down its course, it would have used the end point D as an end and a beginning, using the one point as two; for this reason it must stop, and not simultaneously have come to be at D and have departed from D, since it would be there and not be there in the same now. And ... it is not possible to say that H is at D in a cut in time, but has not come to be at it or departed from it. For it is necessary to arrive at an end that is actively there, not potentially. So while the points in the middle are potentially, this one is actively, and from below is an end, from above a beginning; it therefore belongs in the same way to the motions. It is therefore necessary to something that turns back on a straight line to stop. It is therefore not possible for a motion in a straight line to be everlastingly continuous (*Physics* VIII.viii.262^{a}12-17; [...] 262^{b}23-263^{a}3, Sachs).

There is no mistaking the actual words. Aristotle insists, in no uncertain terms, that for a moving object to reverse its motion, a finite time segment of rest must

---

[8]See Appendix A for objections to the suggestion that according to Aristotle, earth becomes less heavy as it recedes from the center of the cosmos, and heavier as it approaches it.

separate the two stages of motion, always, without exception, and regardless of any other consideration. The unqualified generality of the statement seems to require that when a falling boulder collides with an upward hurled sand pebble and reverses its motion, it must come to a full stop for a brief period.[9] It must do so to respect the pebble's need for this stop so as not to use the end-point of its upward trip as the beginning of its downward trip at one and the same time. Furthermore, since the downward motion following the collision is purely natural, it must accelerate gradually again from no motion, as if the combined boulder and pebble had been released at the collision point to fall from rest. Even a casual observation of phenomena will reveal the inadequacy of such an account, and in his argument for the intervening rest, Aristotle himself appeals to observable phenomena: "The belief that it must stop rests not only on the senses but also on reason." (*Physics*, VIII. viii.262[a]19-20, Sachs). Before attempting to explain this demand by Aristotle, consider the motion again with regard to mover and mobile, according to the principles of motion laid out in levels (1)–(3).

In the case of a volume of rising fire that gradually turns into earth and switches from upward to downward motion, the transition is from one prevailing natural motion to another, forced on one part or the other of the volume under the continuing presence of two opposed original movers. During the entire process the two tendencies—that of fire up, and that of earth down, are in actual competition with one another, each in turn forcing its natural motion upon the other through the mediation of the surrounding medium, the net effect reflecting the difference between their respective intensities. Within the changing bulk, no part is at same time both actual fire and actual earth. The excess of the more powerful of the two actual parts over the weaker impresses effective force on the medium, which moves the weaker of the two against its own nature. Therefore, there is no question here of one motion actively coexisting at any time with its physical contrary, but rather a case of complementary switches between actual and potential as previously explained. The intervening period of rest comes about because part of the continuous transition from excess of fire to excess of earth takes place below the threshold of motion. The dynamics, then, reproduces the requirement of rest separating the two opposed motions, because the threshold of motion is imposed on a continuous change of force. However, no such continuity exists in the collision between the boulder and the pebble, so why should the pebble go through a period of rest?

Consider the quoted passage more carefully: it focuses not on the dynamics of the mobile and its movers, but on the path; not on the efficient aspect of motion, but on a formal aspect. The problem arises not because the mobile is actively moving in two contradictory motions at the same time, but rather because the end-point of the path seems to act at the same time as both an end and a beginning. Aristotle suggests that the act of turning back makes a point that was merely a potential end

---

[9]Galileo refers to this as a "well known" argument against Aristotle. See I.E. Drabkin and S. Drake, (trans.), Galileo Galilei, *On Motion and On Mechanics*, (Madison, Wisconsin, University of Wisconsin Press, 1960), p. 97, or p. 326 in Favaro's critical edition of Galileo's works.

into an active one, and if at the same instant the body also begins its motion back, the point must act as a beginning at the same time. Opposed actualities cannot coexist simultaneously in the same place. Hence, if one point is to serve both as end and beginning, it must do so at different instants of time. Between them neither motion to nor motion back could exist and hence it must be a period of rest.[10]

While the argument as such may seem compelling, one wonders whether it is really so. After all, the end points of the line are just geometrical markers of the limits of a line segment, between which a body oscillates. Limits do not require the further appellation of beginning or end. The back and forth movements belong to the moving body, not of the line as such, and so are the terms beginning and end of the motion. Up to a point in time the body moves actively to the left, potentially to the right, from that point it moves actively to the right, potentially to the left. The end points of the line merely mark the two extremes between which the motion takes place. They never belong to the active motion as beginning and end, so no conflict occurs, neither dynamically, nor formally. The body never arrives at an active end point, which then instantaneously switches role into an active beginning point. It merely passes through a turning point, where its *potential* backward motion becomes *active* backward motion, while the turning point remains a turning point. Indeed, Aristotle himself says:

> But of some things there are not extremities, and of others the extremities are different in kind though like-named; for how could the end of a line and the end of a stroll touch and become one? (*Physics* V.iv.228ᵃ25, Sachs).

It may be objected, therefore, that Aristotle's terminology simply begs the question. The ends of a line are neither beginnings nor ends; they are merely limits. A line, as a geometrical entity, stands outside time, and its ends do not inherently possess the properties "beginning" and "end" that make sense only for a timed process or for a structured logical argument. As for the process of oscillation, Aristotle imposes the terms "beginning" and "end" a priori, while the same points of the process may simply be termed "turning points," and need not be considered as ends or beginnings. Up to, but not including a turning point, the body moves actively to the right, potentially to the left; from the turning point, inclusive, it actively moves left, potentially to the right. No conflict of two opposed activities existing simultaneously occurs. A pendulum begins to swing at the single time and place where its bob was first released; swinging ends when the pendulum's bob finally comes to a stop near the bottom of its trajectory (exact stopping point determined by the threshold of motion). All intermediary limits of swinging are merely turning points, and the pendulum passes through them in a continuously reciprocating process that finds its natural end at the bottom. Aristotle is concerned to avoid endless reciprocating motion along a straight line, but the realization that such motion can never be natural to any terrestrial body already takes care of that: it

---

[10]Aristotle considers his argument against the possibility of un-disrupted reciprocating motion to apply generally, and not as confined to locomotion alone. He makes this explicitly clear in *Physics*, VIII.viii.264ᵇ2-8.

implies a mover of infinite power, the possibility of which Aristotle denies from the beginning. Once again, the threshold of motion will impose brief periods of rest between the swings of a pendulum.[11] But the back and forth motion of a ball, powerfully hit by swinging rackets at its turning points, need not suffer such interruptions.

So far, then, it seems difficult to find compelling reason behind Aristotle's argument. Except where mandated by the threshold of motion, his dynamical principles do not appear to require an absolutely mandatory rest period between the back and forth phases of reciprocating motions. The formal argument hangs on a legitimate, but not compelling a priori appellation of turning points in continuous reciprocating motion "ends" and "beginnings." But the claim itself is unequivocal: periods of rest must separate the phases of all reciprocating locomotion, regardless of dynamical specifics. It may, therefore, be taken as an independent postulate, and there is little room for doubt that Aristotle took it seriously. It may be imposed prior to any dynamical considerations as a prerequisite. In this case, dynamics applies up to the mandatory rest, and following it. Dynamics, however, does not produce the required rest as a dynamical consequence on each and every occasion. Therefore, since the postulate is an independent one, the dynamics developed in levels (1)–(3) operates without contradiction if the postulate is not adopted, and it need not be included for the purpose of developing the dynamics of locomotion along select Aristotelian lines. This, however, comes at the price of abandoning any claim to a plausible reconstruction of Aristotle's own thinking, because it ignores his explicit requirement that a period of rest must break all reciprocating motions, without exception.

There exists, however, another way to approach the difficulty without sacrificing any of Aristotle's requirements, by looking more closely at the medium with the understanding that in all circumstances it always serves as the supplier of effective moving force on any mobile. The effective force must reside in medium layers, because Aristotle considers that all material bodies are three-dimensional. He does not view geometrical surfaces as material bodies and therefore they cannot act as physical movers. This transforms all collisions from instantaneous events into time consuming processes. In particular, let the process of collision begin not when

---

[11]For another example, consider the case of a bucket being gradually filled with water. Its weight grows gradually too, without interruption. Assume it starts out at weight A, ends up at weight C, and passes through weight B. In keeping with Aristotle's analysis in *Physics*, VIII.viii. 263^{b}10-23, the bucket is never in a state of weight B, it merely passes through this state at a cut in time. But now consider the bucket as tied to a rope wrapped round a pulley and connected on the other side to a load of constant weight B. Clearly, before it matches the weight of B, the bucket would move up. This motion should reverse (with the proper lag created by the motion prolonging effect of the air) once it exceeds the weight B. The change in the direction of moving force must be perfectly continuous, because the weight of the bucket grows continuously through B. Despite this, a period of rest will intervene between the resulting switch in the direction of motion, because the continuity of the change in force ensures that some of it must take place below the threshold of motion. Once again, then, the threshold of motion matches the dynamics to the purely formal requirement that rest must separate between the phases of reciprocating motion.

oppositely moving bodies first come into direct contact, but when the medium shells around them begin to overlap. All motion changes that accompany the collision begin to evolve from this point. Figure 3.6 provides a sense of how the process develops, allowing for the mandatory period of rest for the body that reverses its course, while the other body merely slows down as it continues to move in the same direction throughout the process.

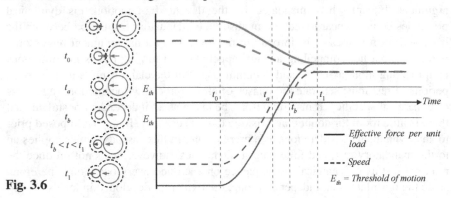

Fig. 3.6

Figure 3.6 sketches the process of collision between two horizontally moving bodies of unequal weight. After the collision, the bodies stick to each other and move as one. Above the time-line, solid and dashed lines mark respectively force and motion to the left; below the time line, solid and dashed lines mark force and motion to the right. Intensities are indicated by distance from the time-line in either direction (force and speed scales should be considered different, and the graphs are superimposed only to give a qualitative sense of the process). The illustrations to the left of the diagram show locations of the colliding bodies at six different instants. Collision begins at t_0, when the two air shells first touch, and ends at t_1, when the surfaces of the moving bodies make physical contact. Only from t_1 on do the two bodies move together at the same speed. Because of the connection between the air shells at t_0, the forces on the moving objects begin to change. They equalize at t_1 when the two bodies make physical contact, and then move on as one body under the effective force in the now unified surrounding shell. The total effective force in the unified air shell just after t_1 is the difference between the effective forces in the two air shells just prior to t_0. In other words, the difference between the initial and final effective force on the small object is the same as the difference between the initial and final effective force on the large object.[12] It follows necessarily that the change in force per unit load on the smaller load is greater than the change per unit load on the larger load, as shown by the solid lines in the diagram. The force per

---

[12]This does not take into account the continuing decay of effective force in the air layers. The decay would be very small as long as the time interval (t_0, t_1) is sufficiently short, but in principle the force in the combined shell just after t_1 is less than or equal to the difference between the forces in the two air shells just prior to t_0.

unit load on the large load never falls below the threshold of motion, hence its speed merely diminishes in proportion to the force until it reaches the final level at t_1. For the small load, as shown by the solid blue curve, the force per unit load changes direction from rightward to leftward, so there must exist t_a and t_b between which the change takes place below the threshold of motion. Therefore, during this time interval the small load must be at rest. The analysis of collisions between vertically moving bodies must take into account the difference between vertical weight and horizontal load, but the process of collision remains essentially the same, with the unavoidable interval of rest separating opposed motions.

In conclusion, medium dynamics underpins, unifies, and gives coherence to Aristotle's various discussions of locomotion. Specifically, the idea that thin medium layers invariably act as the effective movers of all mobiles transforms collisions from instantaneous changes into time-consuming processes. In this framework, the threshold of motion imposes a dynamically mandatory period of rest between reversals of motion in collisions, just as it does in the case of free-flying, upward hurled projectiles. In so doing, it provides the efficient aspect match to Aristotle's formal requirement that a period of rest must intervene between all motion reversals. However, the falling boulder that encounters an upward hurled sand pebble does not have to stop to respect the pebble's need to be at rest between its phases of upward and downward motion. Rather, as the analysis in terms of thin medium shells shows, the boulder will slow down gradually over a brief period of time. During that same period, the pebble's speed will diminish, then grow in reverse to match the boulder's speed, and within the period of change there will always be a shorter period during which the pebble will be at rest.

Heaviness, Lightness, and "the Dense Is Heavy and the Rare Light"

Chapter nine in book four of Aristotle's *Physics* continues a long series of arguments against the possibility of vacuum. Here, Aristotle responds to some of his predecessors who used the assumption of vacuum to explain the phenomena of compression and expansion. Aristotle counters by showing how to explain these without assuming an empty space into which less or more matter could be packed. In this context he introduces an unexpected connection between density and heaviness.

Aristotle never questions the reality of compression and expansion, but instead of taking them to indicate the existence of a vacuum, he regards them as yet another pair of opposite attributes between which their material carrier changes:

> ... our own explanation is based on the established principle that it is the same matter (ὕλη) which experiences the contrasted affections of heat and cold and the other physical opposites, passing either way from a potential to an actual affection, never existing in separation from all attributes, but maintaining its identity through the changing modes of its existence—for instance its colour or its temperature.

Similarly the matter of a body may also remain *identical* when it becomes greater or smaller in bulk. This is manifestly the case; for when water is transformed into air *the same matter, without taking on anything additional*, is transformed from what it was, by passing into the actuality of that which before was only a potentiality to it. And it is just the same when air is transformed into water, the one transition being from smaller to greater bulk, and the other from greater to smaller." (*Physics*, IV.ix, 217ᵃ21-31, Wicksteed and Cornford 1929, my italics).

The transition from water to air involves more than density change, since liquids and gasses differ by other attributes as well. Water vapor, for example, is compressible; water is not. Ignoring these additional differences, the idea seems simple enough: a given quantity of fundamental matter can carry a set of properties, each of which is located within a range specified between two opposites, light-heavy, bright-dark, hot-cold, soft-hard, and now, Aristotle adds, large in volume, and small in volume. The amount of fundamental matter is independent of all such properties, e.g., a given portion of hot matter is independent of its heat—the same portion can be hotter or cooler. In the same manner exactly, a given portion of fundamental matter is independent of the volume it carries. The larger the volume it carries, the rarer it is, and the smaller the volume, the denser it is, but the amount of matter does not change. The same goes for heaviness and lightness: they are essentially tendencies to the center of the universe and away from it, respectively. Weight, then, being an indication of a body's place along the spectrum from the absolutley light to the absolutely heavy, is as such independent of density: rare matter, on this view, can be endowed either with great heaviness or great lightness.

This dissociation of density from weight should be contrasted with the fundamental linkage between density and quantity of matter in the definition that opens Newton's *Principia*. At first sight, they appear similar, because in both greater density means greater quantity of matter in a given volume. In classical mechanics, however, mass and weight are intimately linked: a room containing a single marble of lead has greater density than a room of equal size, containing dozens of equally sized marbles of balsa wood that together weigh less than the single lead marble. Not so Aristotle: weight bears no relation to the quantity of matter. The same matter, at the same place on the earth's surface, bearing the same volume, can carry any weight, from the absolute lightness of fire to the absolute heaviness of earth, and all intermediate combinations. For Newton, then, under similar conditions, the heavier is by definition the denser and vise-versa, in sharp contrast to Aristotle, where the dense and the heavy are by definition independent of each other, which is why the following comes as a surprise:

But the dense is heavy and the rare light. {Yet just as the circumference of a circle when drawn together into a smaller one does not receive some other concave thing, but what was present is drawn together, and every bit of fire whatever that one might take will be hot, so also in general is there a drawing together and pulling apart of the same material.}[13] And there are two aspects each of the dense and the rare, for both the heavy and the hard seem to

---

[13]Wicksteed and Cornford (1929) omitted the bracketed lines in their translation, on suspicion of non-authenticity. Waterfield (1996) did the same.

be dense, and their opposites rare, both the light and the soft. But the heavy and the hard are not in unison in lead and iron (*Physics*, IV.ix. 217^b11-20, Sachs 1995).

Even if the often excised lines 12–16 (in { }) are spurious or misplaced, they are certainly in keeping with the basic idea, providing further illustrations of how any property whatever may be carried in varying degrees by the same substrate, requiring no addition of new material for the sake of changing some property. Most importantly, none of that reflects on the basic connection announced in these lines, namely, the dense is heavy; the rare is light. Considering that heavy is fundamentally that which tends inward (toward the center of the universe), and light is that which fundamentally tends outwards, this association is hardly self-explanatory or intuitive. In fact, it seems to create difficulties.

A large bubble of air floats in water faster than a smaller one.[14] The lightness of the large bubble is therefore more forceful than that of the smaller. It should not matter if the large bubble is compressed to the size of the small one: all the matter is still there, carrying the same ratio of lightness to heaviness. It would seem, therefore, that compression should make the heavy heavier, and the light lighter, but Aristotle clearly says that the dense is heavy, and the rare light. However, the apparent contradiction disappears upon further reflection on how absolute weight, relative weight, and density relate to one another.

Critical to keep in mind for the ensuing discussion is that fire is lighter than any substance containing lightness and heaviness in the sense that a bubble of fire immersed in any other substance will strive to rise to the top, regardless of its size. Likewise, earth is heavier than any substance containing lightness and heaviness. This is not to say, however, that a bubble of fire will rise in some medium faster than a larger bubble of some other substance in the same medium. Specifically, regardless of size, any bubble of fire will rise in air; but in water, a sufficiently large bubble of air would rise faster than a small bubble of fire. Similarly, in air a grain of lead will not fall faster than any ball of wood, and, in air, will not necessarily tip the scale against any ball of wood. Any interpretation of Aristotle proposed in order to settle the apparent difficulty of his statement that the dense is heavy and the rare light must accommodate these phenomena.

For all materials containing heaviness, lightness, and any combination of them, use the definitions in the left box, which will be kept there throughout the ensuing discussion for ease of reference. The weight per unit matter, ω, assessed here by the ratio of heaviness to lightness, $\beta{:}\kappa$,[15] is a way of locating any substance on the

---

[14]As Aristotle says of fire: "… yet the larger quantity {of fire} moves upwards more quickly than the small." (*On the Heavens*, IV.ii.309^b13.

[15]Strictly speaking, this is a "bad" ratio, to the extent that ratio is a relationship between magnitudes of the same kind. In the strict sense, two bodies should be placed on the scale of weight per unit matter by comparing the ratio of their respective heaviness to the ratio of their respective lightness. Namely, if per unit matter $\beta_1{:}\beta_2 > \kappa_1{:}\kappa_2$, then body 1 is heavier than body 2; if $\beta_1{:}\beta_2{::}\kappa_1{:}\kappa_2$, then they have the same weight; and if $\beta_1{:}\beta_2 < \kappa_1{:}\kappa_2$ then body 1 is lighter than body 2. To make the discussion less cumbersome, this more rigorous form is avoided. The same holds for the other mixed ratios used here.

spectrum between the two extremes of absolute lightness and absolute heaviness: the greater the ratio, the heavier the body, or closer in natural place to the natural place of earth. The ratio does not apply to the extremes of pure fire and pure earth, since the ratio of 0 to a finite magnitude does not exist. With all of this in mind, consider the act of compression. It diminishes the volume carried by a given quantity of matter, or, under compression, v—the volume per unit matter diminishes. Compressing by a factor of n ($n < 1$ means rarefaction):

$$V \to \frac{V}{n} \Rightarrow v_n = \frac{v}{n}, \tag{3.1}$$

W	Overall weight
Q	Quantity of matter
V	Volume
$\frac{V}{Q} = v$	Volume per unit matter
$\frac{W}{Q} = \omega$	Weight per unit matter
$\frac{\omega}{v} = \rho$	Weight per unit volume
β	Heaviness per unit matter
κ	Lightness per unit matter

$$\frac{\beta}{\kappa} = \omega = \frac{W}{Q}.$$

where v_n is the volume per unit matter after compression by a factor of n. Compression only changes the volume associated with a given portion of matter, not the heaviness and lightness associated with that same matter and hence not its overall weight, W, that marks its place on the spectrum between the absolutely heavy and absolutely light. For the same reason, however, weight per unit *volume* after compression is:

$$\rho_n = \frac{\omega}{v_n} = n\frac{\omega}{v} = n\rho. \tag{3.2}$$

Again, note that compression does not change the ratio of heaviness to lightness in the compressed matter, so regardless of compression, it still occupies the exact same place on the spectrum between the absolutely heavy and absolutely light.

Weight, however, whether per unit matter or per unit volume, is of little use in actual phenomena that always take place in some medium. Aristotle is very clear on this issue, and in *On the Heavens* Book IV he indicates that for the purpose of investigating what rises and what sinks, it is the weight of a substance *relative* to its surrounding medium that determines how it moves naturally.[16] He also explains that the relative weights of any two substances must be compared in equal volumes of them.[17] In other words, to determine whether a body sinks, floats, or hovers in a given medium, the body's weight must be compared to the weight of an equal volume of the medium. What Aristotle does not do, however, is instruct how mathematically to compare the equal volumes of substance and medium. Archimedes associated the weight of a substance relative to its surrounding medium by the difference between the weight of the substance and an equal volume of the medium. Continuing in these terms therefore goes beyond Aristotle, but as will soon become apparent, not necessarily beyond his physics. Signifying the weight per unit volume of medium by ρ_m, the weight per unit volume of the immersed body by ρ_b, and the body's effective weight in the medium by W_e, Archimedes's measure is:

$$W_e = V(\rho_b - \rho_m). \qquad (3.3)$$

W	Overall weight
Q	Quantity of matter
V	Volume
$\frac{V}{Q} = v$	Volume per unit matter
$\frac{W}{Q} = \omega$	Weight per unit matter
$\frac{\omega}{v} = \mu$	Weight per unit volume
β	Heaviness per unit matter
κ	Lightness per unit matter

$$\frac{\beta}{\kappa} = \omega = \frac{W}{Q}.$$

For $\rho_b > \rho_m$, W_e which measures the body's heaviness relative to the surrounding medium, grows with volume. Therefore according to Aristotle (and every day experience) it will fall faster in the medium. Now compress the body in Aristotle's sense of endowing its matter with a smaller volume. Using Eq. (3.2) with $n > 1$ yields:

---

[16]"In air, for instance, a talent of wood is heavier than a mina of lead, but in water it is lighter." (*On the Heavens*, IV.iv.311^{b}4, Guthrie).

[17]By "light" and "lighter" in a relative sense we mean that one of two bodies of the same size, each possessing {heaviness}, whose natural velocity in a downward direction is exceeded by that of the other." (*On the Heavens*, IV.i.308^{a}30-31, Guthrie).

$$W_{e,n} = \frac{V}{n}\left(\rho_{b,n} - \rho_m\right) = \frac{V}{n}\left(n\rho_b - \rho_m\right)$$
$$= V\rho_b - \frac{V}{n}\rho_m > W_e.$$

(3.4)

As already noted, compression does not change the body's location in the spectrum between absolute heaviness and absolute lightness, yet in a given medium compression does increase its effective weight in comparison to the effective weight of its uncompressed state. An equal volume of the compressed body will therefore fall down faster in the medium then the same volume of its uncompressed state.

To complete the picture, apply this to a body with $\rho_b < \rho_m$ and consider the effect of increasing its volume without altering its density, that is to say, by adding to it more matter carrying the same volume per unit matter. Both its overall weight and its volume will grow proportionately, except that in keeping with Eq. (3.3) "greater weight" now means greater lightness relative to the surrounding medium. It follows, therefore, that a larger bubble of uncompressed air immersed in water has greater lightness relative to its watery surroundings than a smaller one, and will float upwards more quickly, or tend to do so more forcefully if resisted. On the other hand, if the air is compressed, Eq. (3.4) shows that it will become heavier in water, and hence float up more slowly. All of this is fully compatible with Aristotle's concepts of the light, the heavy, and the meaning of compression.

In conclusion, the key to the analysis above resides in understanding the terms "heavy" and "light" throughout the discussion as referring to relative weight and not to absolute weight. Then the idea that the dense is heavy and the rare is light fits nicely with Aristotle's fundamental notions of heaviness and lightness, and their physical significance in terms of natural motion. Furthermore, one does not need to accept Archimedes's rejection of lightness as an absolute quality in order to adopt his measure of relative weight in a material medium, and Aristotle's physics can, therefore, accommodate Archimedes's law of buoyancy with no ensuing internal conflicts.

The Geometry of Heaviness: On Projectiles that Reach Peak Speed at Midflight

Aristotle's discussion in *On the Heavens*, II.6 sets out to show that the motion of the first heavenly sphere is a pure uniform rotation, but it opens with a very brief classification of non-uniform motions, and an explicit reference to the terrestrial phenomenon of thrown objects:

> If it moves irregularly, there will clearly be an acceleration, climax, and retardation of its motion, since all irregular motion has retardation acceleration, and climax. The climax may be either at the source or at the goal or in the middle of the motion; thus we might say that

for things moving naturally it is at the goal, for things moving contrary to nature it is at the source, **and for things whose motion is that of a missile it is in the middle**. (τοῖς δὲ ῥιπτουμένοις ἀνὰ μέσον). (*On the Heavens*, II.vi:288ª17-22, Guthrie 1939).[18]

The emphasized line, clearly at odds with intuition and every day experience, also seems irreconcilable with Aristotle's own theory of locomotion, which usually identifies projectiles as an instance of violent motion. Taken uncritically under this view, projectiles should generally have their greatest speed in the beginning—not the middle—of their motion. Considered in this light, the trail of puzzled comments from ancient to modern times seems hardly surprising.

The first thing to note about Aristotle's comment on non-uniform motion is that it consists of a general observation followed by three brief illustrative examples. The general observation states that all non-uniform motion must have acceleration, peak, and deceleration. That "peak" (ἀκμή) here must refer to the speed of motion and not to position is made amply clear by the statement that for things moving naturally the peak is at the goal, which for heavy bodies is at the bottom of the trajectory. To further illustrate the general observation, there follow references to the basic categories of natural and counter-natural motions, and to an aspect of projectile motion. Naturally, the explicit reference to projectiles tends to direct the mind to practical experience with slings, catapults, bows, and similar devices. The context, however, strongly suggests that precisely such devices should not be associated with the projectiles relevant to this discussion.

The key to the resolution of Aristotle's observation on projectiles that attain top speed at mid-motion is in *On the Heavens*, Book II.4, which explains the reason for the sphericity of water surfaces as follows:

As for the sphericity of the surface of water, that is demonstrable from the premise that it always runs together into the hollowest place, and 'hollowest' means 'nearest to the centre'. Let AB and AC be straight lines drawn from the centre, and joined by BC. Then the line AD, drawn as far as the base, is shorter than the original lines from the centre, and the place occupied by D is a hollow. Hence the water will flow round it until it has filled it up. The line AE, on the other hand is equal to the radii, and thus it is at the extremities of the radii that the water must lie: there it will be at rest. But the line which is at the extremities of the radii is the circumference of a circle. Therefore the surface of water; represented by BEC, is spherical. (*On the Heavens*, II.4: 287ᵇ5-14, Guthrie 1939)

---

[18]Beyond this short remark there is no further analysis of non-uniform motion in the chapter, so the observations on acceleration, deceleration, and climax serve merely as brief general reminders, not as analytical investigations in their own right. The entire chapter focuses on showing that the circular motion of the ethereal spheres is necessarily uniform. The introductory statement that sets out the subject for discussion notes parenthetically that this uniform motion is apparent only in the first sphere that provides the daily rotation common to all heavenly motions, and that below it the motion is already a combination of several components. This seems to reflect the Eudoxan homocentric structures, which should be kept in mind as background to the entire chapter.

Thus the idea proposed by Leo Elders, that the word ἀκμή does not denote "a state of highest intensity" but "the highest point of the trajectory" is unacceptable; Leo Elders, *Aristotle's Cosmology* (Assen 1966), p. 210).

The indicated diagram, with embellishments marked in small italicized letters, should look more or less like Fig. 3.7.

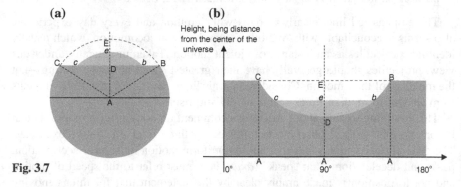

Fig. 3.7

As described in the text, the diagram clearly intends to represent a plane projection of a three dimensional figure, where the circle represents the sphere of the earth whose center, A, coincides with the center of the universe. BDC then represent the circular plane left after the dome represented by BEC has been sliced off the earth.[19] The text explains that for water, the midpoint D between B and C signifies the bottom of a cavity. Therefore, any quantity of water will collect there in the form of domes, like *bec* in Fig. 3.7a, sliced from spheres centered on A. When completely filling the cavity, the water surface will coincide with dome BEC, because it is everywhere equally distant from the center.

Taking "gravity" to stand for the tendency of all heavy things toward the center of the universe, Aristotle's discussion creates a distinction between "flat" in a visual context, and "horizontal" in a gravitational context. The visually flat is a geometrical plane, such as the one projected into line BDC in Fig. 3.7a. For heavy objects, however, Aristotle calls the same plane "a hollow," and the sphericity of water surfaces reveals that "horizontal" for heavy objects is the surface that is everywhere equidistant from the center of all heavy things, namely point A—the center of the universe, with which the earth' center coincides. Thus, in the diagram, the visually spherical surface BEC, being everywhere equidistant from A, is gravitationally horizontal, while the visually flat plane BDC is a gravitational cup, bottoming out at D (compare Fig. 3.7b, where distance from A is graphed around the upper half of the circle that represents the earth). To the extent that "heavy" means that which seeks the center of the universe, a heavy ball left to its own devices just slightly right of C would roll under its own weight toward the bottom of the cup at D, even though visually it would seem to be rolling along a flat plane.

---

[19]The sphere around A need not be the size of the earth. It can be much larger, or much smaller. What must remain in all cases is that the circular arc subtended by points B and C must be a significant part of the sphere's circumference.

We may now return to the remark on projectiles that attain maximum speed at mid-motion. In a comment added to his own translation of *On the Heavens*, W.K.C. Guthrie wrote:

> Surely one would naturally suppose this to mean the middle point of the trajectory.... Then comes the obvious difficulty that this is contrary to fact. At the middle point of its trajectory a stone from a sling is moving at its slowest. I do not see how this difficulty is to be avoided if the text is what A. wrote.[20]

However, consider the possibility that just as the discussion pertaining to the sphericity of water surfaces does not seek observational confirmation in swimming pools, the projectile in our text does not refer to stones launched by sling-shots. In both cases the scope is cosmological, in keeping with the nature of *On the Heavens* as a whole, and especially so in the first two books. In other words, Aristotle's remark here should not be read as part of a theoretical treatise on the ballistics of slings, catapults, and bows. It should, rather, be read as a fundamental observation on the same global scale that reveals the sphericity of water-surfaces (see Fig. 3.8).

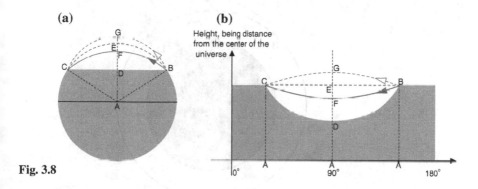

Fig. 3.8

Using for reference the figure that Aristotle suggested in arguing for the sphericity of water surfaces, the heavy object that flies from B to C along BFC seems, visually, to trace a convex path (Fig. 3.8a). However, as Aristotle clearly explains with respect to water surfaces, in the geometry of heaviness "down" means reducing the distance to the center of the universe, A. In this geometry, as Fig. 3.8b shows, the gravitationally horizontal path from B to C is the visually circular arc BEC that maintains constant distance from the center A. Therefore, to a heavy projectile the flight path BFC is concave, because it approaches the center from B to F, and recedes from it from F to C. Aristotle's theory of motion therefore requires that on all such

---

[20]*On the Heavens*, tr. W. K. C. Guthrie, pp. 170–171, n. a.

trajectories the projectile must reach peak speed at mid-motion, just as his text actually maintains. Commonly observed projectiles slow down at midflight because like BGC in Fig. 3.8 their trajectories invariably lie above the circular arc BEC.

The implied classification, then, falls into three categories of heavy locomotion (see Fig. 3.9). Category (1) contains motions that have a downward component, reach top speed at the end of the motion (and contain, most prominently, purely natural fall). Motions that do not attain top speed at the end split into two further categories. Category (2) contains forced motions that follow trajectories above the circular arc drawn about the center through the launching point, and have top speed at the beginning. Category (3) contains forced motions that follow a trajectory below the circular arc drawn about the center through the launching point, and reach top speed at mid-motion.

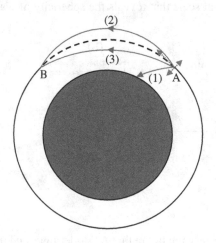

Fig. 3.9

Aristotle's text, then, suggests a reasonable distinction between types of projectile motion under this three-headed classification. Since clearly not all projectiles reach top speed at mid-motion one may complain about the confusion created by the unqualified use of "projectile" in reference to the special class of those moving objects that obtain maximum speed in the middle. The nature of the observation is, however, that of a brief remark in a chapter that is dedicated to the study of circular motion. Aristotle's intention here was not to develop the remark into a detailed study of terrestrial projectiles, but to single out an unintuitive, but highly instructive class of motions without which any classification of projectile motion would remain incomplete. For the same reason one could hardly fault Aristotle for failing to discuss the borderline case of projectiles launched tangentially to the circular arc through the launching point, and for failing to take into account the resistance and decaying dynamical effect of the medium.[21]

---

[21]In the case of downward hurled heavy projectiles, for example, the dual action of the medium in Aristotle's theory requires important distinctions. When hurled at a speed below the terminal speed of its natural fall, the peak speed will occur either at the end, or at the point of travel where it reaches terminal speed. When hurled at terminal speed, the projectile will travel downward at a constant speed throughout the duration of its journey. When hurled at a speed greater than its terminal speed, the projectile's motion will invariably start at peak speed, decaying gradually into its terminal speed. Medium resistance determines the terminal speed of a given projectile. All the gradual speed transitions in the various cases of projectile flight are direct consequences of the motion prolonging effect of the medium in accordance with the discussion in *Physics* VIII.x.

Chapter 4
The Dynamics of Balance: The Winch and the Lever in the Pseudo Aristotelian *Mechanical Problems*

How to Balance a Winch Using Aristotle's *Physics*, VII.V

The object of investigation so far has been Aristotle's theory of inert matter in motion, and therefore, the emphasis has been on the dynamical relationships between active locomotion, force and resistive effects that become manifest only in the presence of active motion. However, the basic dynamical principles between force, motion, and load can also be used to account for balanced rest, as the following will show.

For present purposes, a winch consists of two cylindrical blocks of different radii, glued to each other so as always to rotate at the same angular speed around their common central axis. In Fig. 4.1, the winch is horizontal, while the two heavy weights hang vertically from wires that connect to the winch via pulleys. Given the winch's two radii, the problem is to determine the weight ratio required for balance to obtain.

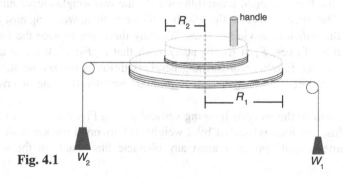

Fig. 4.1 W_2 W_1

The solution described below has no historical pretense—it does not follow a discussion in Aristotle or in any other extant Greek source. The purpose is to show how Aristotle's theory of locomotion may be applied to derive the balance

© The Author(s) 2015
I. Yavetz, *Bodies and Media*, SpringerBriefs in History of Science and Technology, DOI 10.1007/978-3-319-21263-0_4

condition. This will help put in sharper relief the discussion of the winch and the balance in the first problem of the Pseudo-Aristotelian *Mechanical Problems.*[1]

As a first step toward the solution, ignore the hanging weights in the diagram above, and consider the horizontal winch as connected by the larger wheel to W_1 alone, lying with the winch on the same horizontal surface with negligible frictional resistance. Apply force of some arbitrary intensity to a handle set at some arbitrarily chosen point on the winch to rotate it clockwise and reel in W_1, and suppose that as a result W_1 moves at speed V_1. Since the smaller wheel is glued to the larger and rotates with it, the speed, V_2, of any point on its rim is uniquely determined according to $V_2:V_1::R_2:R_1$. Now, according to *Physics*, VII.v., if two loads, W_1 and W_2, are moved by the same force, their speeds relate inversely as the loads, namely: $W_1:W_2::V_2:V_1$. In our horizontal winch, however, once V_1 is fixed, so is V_2. Consequently, if we wish to apply the same force to the same handle but in the opposite direction, to reel in W_2 at the V_2 that maintains the original V_1 (but in the opposite direction) at the rim of the larger wheel, we must require:

$$W_1 : W_2 :: V_2 : V_1 :: R_2 : R_1,$$

Or, in more familiar form:

$$\frac{W_1}{W_2} = \frac{V_2}{V_1} = \frac{R_2}{R_1}.$$

The first proportion reflects Aristotle's general law, while the second reflects the particular constraints due to the geometry of the winch. Therefore, choosing the weights such that $W_1:W_2::R_2:R_1$ insures that equal and opposite forces will reel the weights in with speeds $V_1:V_2::R_1:R_2$, as required by the geometry of the winch. Naturally, if equal and opposite forces apply simultaneously to the same handle on the winch, they will balance each other out, and the winch will not move at all.

Now, to the lines wrapped around the winch, the two weights represent loads to be moved. The force applied to the winch to rotate it, then, works against the loads applied to the winch rims via the lines. We may therefore, replace the weights by any two pulling forces, F_1 at R_1, F_2 at R_2, such that $F_1:F_2::W_1:W_2$, and the winch will remain at rest. In other words, the general condition of balance for the winch is $F_1:F_2::R_2:R_1$, where F_1, F_2 are the forces applied respectively to the two rims of the winch.

Turning now to the weights hanging vertically as in Fig. 4.1, we must take care not to confuse the load presented by a weight to horizontal motion with the force that the same weight applies against any obstacle that stands in the way of its

---

[1]To Galileo and his contemporaries, the *Mechanical Problems* was an integral part of Aristotle's works. For Galileo specifically, investigation of simple machines provided crucial guidance in his quest for a theory of locomotion (see Valleriani (2010), *Galileo Engineer*, Springer, 2010). Serious doubts with regard to Aristotle's authorship of the *Mechanical Problems* emerged and became widely established only during the 19th century (see McLaughlin (2013)).

downward natural motion. As already shown at length, Aristotle's treatment of heaviness requires a careful distinction between vertical and horizontal motion.[2] The result obtained so far cannot apply directly to the situation depicted in the diagram, because the weights are now made to move vertically, reflecting the effect of their natural downward tendency, *i.e.*, their heaviness. However, while the heaviness of a body is not a force that moves the body downward, it does exert a force against obstacles in its way. To the extent that both the load in horizontal motion and the obstacle opposing force of a heavy object are proportional to its intrinsic natural heaviness, the above result remains in effect. The forces applied by the weights against each other via the winch are to each other as $W_1:W_2$, and hence if we select the weights such that $W_1:W_2::R_2:R_1$, the winch should be balanced, and not move. A confirmation of this in practical tests would verify that loads and downward forces of heavy bodies are indeed proportional, and to the extent that one of them is proportional to the heaviness, so also is the other.

If the winch is now repositioned in a vertical plane, the path from it to the case of a balance with unequal arms is straightforward (Fig. 4.2):

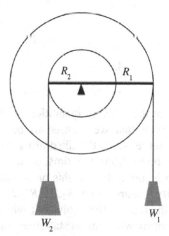

Fig. 4.2 W_2 W_1

Namely, if the pivotal point along the central line of a single horizontal beam divides it in the same ratio as R_1 and R_2 of the winch, with W_1 and W_2 hanging at the respective ends, balance will apply just the same. Furthermore, by drawing the

---

[2]There are simple ways to observe the difference in everyday phenomena: a small weight hanging over the edge of a horizontal surface will descend while pulling a weight much larger than itself along the horizontal surface, but will not be able to do so were the larger weight hanging against it on the other side of a pulley. Less intuitive but equally compelling is that should a weight moving horizontally at a given speed encounter a spring, it will compress it to a lesser degree than it would upon encountering the same spring at the same speed in the course of freely falling downward. If, on the other hand, the weight encounters the spring at the same speed while moving upward, it will compress it less than it does in the horizontal setting.

equivalent winch, it may easily be shown that the orientation of the balance does not matter. As long as the proportion $W_1:W_2::R_2:R_1$ holds, the beam will remain balanced at any arbitrary angle (Fig. 4.3):

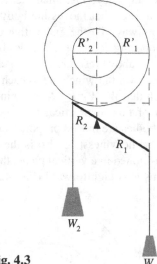

Fig. 4.3

The tilt alters none of the relevant quantities, so $W_1:W_2:R_2:R_1$ remains in effect. Now, however, the winch that we showed to be the equivalent of a horizontal balance can no longer serve, since the tilted arms no longer coincide with its rims, and weights must be hung from the rims of a winch. The new orientation is therefore equivalent to hanging the weights on a reduced winch, with radii R'_1 and R'_2. Triangle similarity ensures that $R_2:R_1::R'_2:R'_1$, and hence $W_1:W_2::R'_2:R'_1$. Therefore, shortening the radii of the equivalent winch according to the tilt does not alter the ratios, so the arms will remain stationary in any tilted position as long as $W_1:W_2:R_2:R_1$ holds.

As stated at the beginning, this derivation of the balance condition for winches and levers is not based on any known Ancient Greek historical document. It merely shows how the balance condition could be obtained from a consistent application of the principles of Aristotle's theory of locomotion, as explicitly formulated in his extant texts. While this does not in any way suggest that someone in Aristotle's time, or after him, must have done something like that, it is at least not outside plausibility that something along these lines could have been done. What actually had been done, is impossible to say with reasonable confidence, considering the meager evidence that survived into our hands.

Putting aside any historical claims on behalf of this exercise, then, it does underscore one important lesson: if Aristotle's physical principles were to be used for the purpose of obtaining the balance law, the argument would almost certainly be dynamical. In other words, the condition would arise from analyzing force and motion, and not from pre-postulated principles of static forces, or from static symmetry considerations like Archimedes's.

The Winch in the Pseudo-aristotelian *Mechanical Problems*

While Aristotle does not provide an explicit formulation of either the law of the lever, or the equillibrium condition for a winch, or for equilibrium in a balance with unequal arms, the pseudo Aristotelian *Mechamical Problems* contains an explicit general statement of the law:

> WHY is it that, as has been remarked at the beginning of this treatise, exercise of little force raises great weights with the help of a lever, in spite of the added weight of the lever; whereas the less heavy a weight is, the easier it is to move, and the weight is less without the lever? Does the reason lie in the fact that the lever acts like the beam of a balance with the cord attached below and divided into two unequal parts? The fulcrum, then, takes the place of the cord, for both remain at rest and act as the centre. Now since a longer radius moves more quickly than a shorter one under pressure of an equal weight; and since the lever requires three elements, viz. the fulcrum—corresponding to the cord of a balance and forming the centre—and two weights, that exerted by the person using the lever and the weight which is to be moved; this being so, *as the weight moved is to the weight moving it, so, inversely, is the length of the arm bearing the weight to the length of the arm nearer to the power.* [my italics] The further one is from the fulcrum, the more easily will one raise the weight; the reason being that which has already been stated, namely, that a longer radius describes a larger circle (*Mechanical Problems*, III.850ª30-850ʰ3, Forster).[3]

The passage refers back to the first question in the treatise, which may create the expectation to find therein a proof of some sort. Unfortunately, the discussion in problem 1 is not easily followed, and does not appear to lead by deductive procedure to an explicit statement of the law that the italicized lines in the above quotation formulate so clearly.[4] The reasoning offered in problem 1 differs markedly from the Aristotellian exercise shown above, and does not explicitly apply the principles developed by Aristotle in the *Physics* and *On the Heavens*. What it does share with Aristotle is the dynamical approach to balance. Namely here, as well as in the Aristotelian exercise above, the examination follows the forced motion of a winch, or a lever hinged at one end. Discussion begins with a question:

---

[3] *"The weight getting moved to the weight moving is the opposite of length to length. And always, the farther from the fulcrum, the easier it will move.* The reason is the aforesaid, that the more distant from the center scribes the larger circle. So by the same force, the mover will manage more the farther from the fulcrum." (Aristotle (2007), pp. 10–11, my italics).

[4] As Joyce van Leeuwen reasonably suggested in conversation, the reference is probably not to the mathematical analysis in the later parts of problem 1, but to the very beginning of the treatise, where the author wonders how a force that cannot on its own move a given load, can do so when applied through a mechanical device.

First, then, a question arises as to what takes place in the case of the balance. Why are larger balances more accurate than smaller? (*Mechanical Problems*, I.848[b]1-2, Forester).

This is quickly broken down into subsidiary problems, the first of which is:

...why is it that the radius which extends further from the centre is displaced quicker than the smaller radius, when the near radius is moved by the same force? (*Mechanical Problems*, I.848[b]2-5)

The answer given at this point is only an indication: "The reason of this is that the radius undergoes two displacements" (*Mechanical Problems*, I.848[b]10, Forester), and now follows a digression that produces the first positive result:

Now if the two displacements of a body are in any fixed proportion, the resulting displacement must necessarily be a straight line, and this line is the diagonal of the figure, made by the lines drawn in this proportion. (*Mechanical Problems*, I.848[b]11-14, Forester)

To see how this comes about, the author suggests looking at the parallelogram ABHΓ (Fig. 4.4):

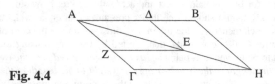

Fig. 4.4

The text follows the motion of a point that starts out at A and moves to B while the entire line AB moves to ΓH with its end points traveling along AΓ and BH, keeping AB parallel to ΓH at all time. This means that maintaining the specified proportion to AB:AΓ while A moves to Δ and line AB moves to ZE requires AB:AΓ::AΔ:AZ. Since the parallelism of ZE and ΓH implies equality of the angles AZE and AΓH, and the respective sides are proportional by assumption, triangles AZE and AΓH are similar. Therefore the movement will be along the diagonal AH throughout the motion. This, then, shows that when a body moves with two unchanging simultaneous motions (unchanging both in speed and direction), then its actual motion will be along a straight line according to the parallelogram construction above.

The text immediately continues to observe that conversely, the motion along a diagonal can always be decomposed into two simultaneous motions. The importance of this to anyone with a smattering of geometrical knowledge is that the converse statements are not symmetrical twins: in parallelogram ABHΓ, only a single diagonal AH can be drawn; on the other hand, the line segment AH could serve as the diagonal of an infinite number of parallelograms. Any straight uniform motion can, therefore, be decomposed in infinitely many ways into two simultaneous uniform motions.

The discussion concludes with the generalization: "if the two displacements do not maintain any proportion during any interval, a curve is produced." (*Mechanical*

Problems, I.848*b*34, Forester). This is demonstrated by reduction to absurd: assume the motion to be a straight line, and complete a parallelogram around it. Since the line is assumed straight, the ratio of the sides will remain the same at all points along the line as previously demonstrated. But this contradicts the initial requirement that the magnitudes of the two component displacements do not maintain a fixed ratio while their direction remains unchanged. Therefore the assumption that the resulting motion is straight must be rejected, and the path must be curved.[5]

Altogether, the discussion regarding the combination of motions in *Mechanical Problems* is undoubtedly one of its most important elements. It is certainly not the first indication in Greek literature of such combinations. Plato, in the *Timaeus* considers the combination of two rotations in mutually inclined planes. In Aristotle's *Metaphysics*, such combinations are multiplied several times over in the account of the homocentric systems of spheres that move the planets. Both of these discussions probably reflect the work of Eudoxos of Cnidos, to whom Aristotle ascribes the first set of homocentric planetary systems that he discusses. Aristotle also very clearly notes that motion upward can never be cancelled by motion sideways. However, these discussions do not teach the procedure of combining motions, and that is where the *Mechanical Problems* stands out. Not only does it describe the parallelogram rule, but it goes on to indicate how it may be used in the case of non-uniform motion. As we are about to see, the motions become closely associated with moving forces that operate in different directions, so the procedure is not confined to kinematics only. If, as the general opinion suggests, the work was composed by an early disciple or disciples of Aristotle,[6] then this indicates that young contemporaries of Aristotle already knew how to add and subtract magnitudes that do not coincide in direction, and even Aristotle himself might well have been aware of the procedure.

Armed with the principle of combining directed quantities, the text proceeds to examine motion in a circle by decomposing it into two mutually perpendicular components: one along a diameter, the other along a line perpendicular to that diameter. Based on what had already been demonstrated, the author immediately

---

[5]The asymmetry of the converse statements actually needs a slightly more careful and better qualified statement. Straight line motion can, at any time, be decomposed into components that maintain the same two directions in space, and the same ratio of magnitudes. Motion along a quarter of a circle, however, can be obtained by two components whose magnitude remains unchanged while they spin about their common point by the same angle in any time segment (not necessarily at constant rate). It can also be generated by two components along the same spatial directions whose magnitudes do not retain constant ratio, which is the case actually shown in the text. So to be more secure the text should have stated that if the components, while retaining fixed directions, do not retain proportionality in magnitude at all times, the resulting motion cannot be straight.

[6]At least one recent translator challenged this view (Aristotle 2007). I am not convinced by his suggestion that the work should be attributed to an "intellectual giant," and that the only choice is Archytas of Tarentum (p. vi). However, his arguments against associating the work specifically with Aristotle's early disciples and the possibility that parts of it could have been composed earlier deserve serious consideration.

observes that one of the motions must be more hindered than the other. In other words, at least one of the component motions and the force behind it cannot be uniform (Fig. 4.5):

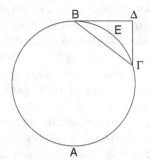

Fig. 4.5

Let ABΓ be a circle, and let the point B at the summit be displaced to Δ by one force, and come eventually to Γ by the other force. If then it were moved in the proportion of BΔ to ΔΓ, it would move along the diagonal BΓ. But in the present case, as it is moved in no such proportion, it moves along the curve BEΓ. And, if one of two displacements caused by the same forces is more interfered with and the other less, it is reasonable to suppose that the motion more interfered with will be slower than the motion less interfered with; (*Mechanical Problems*, I.849^{a}3-10, Forester)

At this point starts the main argument, which is best quoted in full, since its later parts add important clarifications to the earlier ones. The purpose here is to show in detail how the greater interference indicated at the end of the previous quotation works out in the case of circular motion:

For on account of the extremity of the lesser radius being nearer the stationary centre than that of the greater, being as it were pulled in a contrary direction, towards the middle, the extremity of the lesser moves more slowly. This is the case with every radius, and it moves in a curve, naturally along the tangent, and unnaturally towards the centre. And the lesser radius is always moved more in respect of its unnatural motion; for being nearer to the retarding centre it is more constrained. And that the less of two radii having the same centre is moved more than the greater in respect of the unnatural motion is plain from what follows.

Let BΓEΔ be a circle, and XKMΞ another smaller circle within it, both having the same centre A, and let the diameters be drawn, ΓΔ and BE in the large circle, and MX and NΞ in the small; and let the rectangle ΔPΓ be completed. If the radius AB comes back to the same position from which it started, i.e. to AB, it is plain that it moved towards itself; and likewise AX will come to AX. But AX moves more slowly than AB, as has been stated, because the interference is greater and AX is more retarded. Now let AΘH be drawn, and from Θ a perpendicular upon AB within the circle, ΘZ; and, further, from Θ let ΘΩ be drawn parallel to AB, and ΩY and HK perpendiculars on AB; then ΩY and ΘZ are equal. Therefore BY is less than XZ; for in unequal circles equal straight lines drawn perpendicular to the diameter cut off smaller portions of the diameter in the greater circles; ΩY and ΘZ being equal (Fig. 4.6).

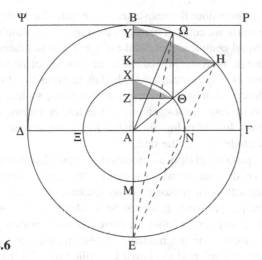

Fig. 4.6

Now the radius AΘ describes the arc XΘ in the same time as the extremity of the radius
BA has described an arc greater than BΩ in the greater circle; for the natural displacement
is equal and the unnatural less, BY being less than XZ. Whereas they ought to be in
proportion, the two natural motions in the same ratio to each other as the two unnatural
motions. (*Mechanical Problems*, I.849ª26-ᵇ5, Forster).

Before all else, note that with the previous discussion about the greater interference coming from motion toward the center, the concluding lines disclose a new
principle: if the requirement is that unnatural and natural components bear the same
ratio, then by implication, the hindrance to motion along a circular arc is evaluated
by the ratio of the unnatural component to the natural one. This comes from
nowhere in Aristotle, and seems to be a ground rule from which all other conclusions in this discussion must follow.

The discussion, however, is puzzling for several reasons. To begin with, the
equilibrium condition for winches and for balances of unequal arms, if anywhere in
this discussion, is truly well hidden. The fundamental inverse proportion between
radii and forces applied to their ends does not appear at any point. Certainly not in
the clear and explicit form found in Problem 3. Perhaps, then, the discussion should
not be taken as a convoluted, and in the end unsuccessful, attempt at getting to the
law of the lever. Perhaps all it wishes to say is that when a force is applied to a
hinged radius, it makes a longer radius generate a greater displacement in the same
time. This does indeed get explicit expression in the observation immediately
following the above analysis:

So that the reason why the point further from the centre is moved quicker by the same force,
and the greater radius describes the greater circle, is plain from what has been said.
(*Mechanical problems*, I.840ᵇ20-21, Forseter)

However, this still fails to streamline the discussion, which uses the requirement
that equal hindrance means equal ratios of natural and unnatural components to

draw a specific conclusion: the resistance is the same for both radii, the text states, provided these ratios are equal. Therefore, the same force should move the two radii so as to cover equal central angles, as the diagram too indicates very clearly. If this is indeed the intent, then the law of the lever should not be sought here anymore, because it is now clearly contradicted. The law requires that to effect equal angular rotations, the forces should be inversely as the radii, so that a smaller force will effect the motion along the longer arc of the longer radius. An equal force will actually swing the longer radius through a larger angle than it would the smaller radius, directly contradicting the text.

All of these puzzles disappear, however, provided two things are done: (1) a careful evaluation of what the writer meant by "the same force;" (2) the assumption is made that regardless of personal idiosyncrasies, the writer was an Aristotelian disciple. Regarding (1), one must not be mislead by the general flavor of the discussion. This is neither a treatise on thrown objects, nor on planetary movement in circles. It is a treatise on mechanical devices, and they constantly guide analysis even if not explicitly referred to. Force is applied to a lever or to a winch at one place, in order to move something that sits somewhere else. Both the fulcrum and the location of the load may change, but the moving force remains at one place. "Same force," therefore, should be taken to imply same force at the same place. So, if a force applied anywhere along the short radius in the diagram is to experience the same resistance when applied at the same distance from the center to the longer radius, the longer radius must move at the same time through the same angular displacement.

This still does not produce the law of the lever. Now, however, assume (2): the writer is well versed with Aristotle's general discussion of movers, mobiles, displacement and time, as in *Physics* VII.v. The equal resistance clause led to the conclusion that the same force at the same place would move the two radii such that during the same time period, their displacements along the circular arcs will be as their radii. Bringing in Aristotle adds the general observation that if at the same time the same force moves two loads such that their displacements are as $R:r$, then the loads must be in the inverse proportion, namely $R:r::W_r:W_R$. In other words, W_R at R and W_r at r on the opposite side of the stationary center would exactly balance one another, because the same force moves them by the same angular displacement. But since equal angular displacement is how winches and balances move, this is clearly the condition for equilibrium to obtain.

It may be objected that bringing Aristotle into the discussion renders the measure of hindrance in circular motion superfluous, for, as we saw, the equilibrium condition may be obtained without it. And anyway, Aristotle is not mentioned anywhere in the *Mechanical Problems*. Taking only the information explicitly given in Problem I without the Aristotelian hints, there is no way to extract the balance condition from the discussions of problem I. In that case, the work merely shows by Problem III, that the writer was well aware of the law of the lever, and had no pretense to derive it in problem I.

Considering, however, that most current historians believe that the work was composed within the early Aristotelian tradition, bringing in Aristotle is not unjustified. True, the resistance measure appears artificial and unjustified to the modern reader, and it is logically superfluous once Aristotle's general laws of motion are postulated. However, this merely shows that the writer, being a creative thinker in his own right, was fascinated by the realization that the measure of hindrance to motion in a circular arc turns out qualitatively compatible with Aristotle's general laws of material bodies in locomotion. As for not mentioning Aristotle—to a close disciple this may not be necessary, being the common ground accepted by his colleagues.

Since I cannot find any reason to take one of these alternatives as overwhelmingly more plausible than the other, I would rather leave this issue unresolved. What now becomes even more clearly emphasized is that regardless of which alternative is chosen, the dynamic approach to the lever and the winch remains unshaken: the discussion examines forces in motion. If equilibrium conditions emerge, they do so based on a dynamical analysis, and not from assumptions regarding static forces and static symmetry conditions.

Chapter 5
Hipparchus on the Theory of Prolonged Motion

Hipparchus, on the other hand, in his work entitled *On Bodies Carried Down by Their Weight* declares that in the case of earth thrown upward it is the projecting force that is the cause of the upward motion, so long as the projecting force overpowers the downward tendency of the projectile, and that to the extent that this projecting force predominates, the object moves more swiftly upwards; then, as this force is diminished (1) the upward motion proceeds but no longer at the same rate, (2) the body moves downward under the influence of its own internal impulse, even though the original projecting force lingers in some measure, and (3) as this force continues to diminish the object moves downward more swiftly, and most swiftly when this force is entirely lost.

Now Hipparchus asserts that the same cause operates in the case of bodies let fall from above. For, he says, the force which held them back remains with them up to a certain point which accounts for the slower movement at the start of the fall. (Simplicius, *Commentary on Aristotle's "De Caelo,"* Heiberg 264.20 ff, in Cohen and Drabkin, *A Sourcebook of Greek Science*, p. 209).

Most modern scholars see Simplicius's account as the earliest documentation of the theory of in-body impetus, which, in the Middle Ages gained ascendency over Aristotle's medium dynamics of motion. Strictly speaking, the account does not explicitly state that the temporary moving force resides in the body. Dijksteruis in 1924, and most recently Michael Wolff,[1] suggest that Hipparchus was actually elaborating Aristotle's medium theory, and that there is no reason to suppose that he anticipated Philoponus with the idea of an "impetus" transferred into the moving body itself. If this is indeed the case, then Simplicius's account provides the best historically documented evidence for the central role of the medium in Aristotle's dynamics of motion and for its influence outside the immediate circle of dedicated Aristotelians. Note, however, that while Simplicius does not describe the force as residing in the body, he also does not say that it resides in the surrounding medium. Simplicius mentions Hipparchus in the context of alternative theories to Aristotle's account of natural acceleration. While Simplicius also says of Hipparchus that contrary to Aristotle he believed bodies to increase in heaviness as they recede from the center, this could not possibly explain the acceleration of falling bodies in everyday experience. Considering the size of the earth's radius, the variation of distance from the earth's center for terrestrial falling bodies is utterly negligible, and

---

[1]Wolff 1987, esp.100–104.

© The Author(s) 2015
I. Yavetz, *Bodies and Media*, SpringerBriefs in History of Science and Technology, DOI 10.1007/978-3-319-21263-0_5

with it any effect it might have on their speed of motion. It would be highly implausible to suggest that Hipparchus was not aware of this, even if Eratosthenes already made the earth's radius half of what Aristotle believed. Simplicius apparently thought that Hipparchus's explanation of natural acceleration is opposed to Aristotle's, which has little or nothing to do with distance from the center, and everything to do with the surrounding air's interaction with natural fall. The remaining candidates for Hipparchus's disapproval are (1) Aristotle's idea that the decaying effective force resides in the medium; (2) Aristotle's unqualified requirement that all reciprocating motion requires a rest period between its opposed motions; (3) Aristotle's inverse proportionality of speed and medium resistance.[2] So while I see no justification for unequivocally attributing the basic idea of in-body impetus to Hipparchus, it is still a perfectly plausible historical possibility that Hipparchus did anticipate Philoponus. One way or the other, nothing in Philoponus comes close to the clever way in which Hipparchus accounts for the stages of flight of upward hurled projectiles.

Galileo's suggestion to consider this as analogical to heating and cooling is very useful.[3] When first placed in a fire, a cold piece of iron does not immediately become hot, but heats up gradually. Likewise, when an original mover begins actively to apply its power to a resting mobile, it does not instantaneously build effective force either in the surrounding medium, or inside the mobile as in-body impetus. Just as in the case of heating, the effective force builds up gradually. That the motion of bodies starts out at rest and gradually accelerates when an external mover applies to them, suggests that the effective mover is not the original mover's power, but the effective force that resides either in the medium surrounding the mobile or inside it, but not in the original mover. Left in a fire for a long time, a piece of iron does not become infinitely hot. Rather, it attains some terminal heat that it does not exceed. It keeps the terminal heat constant so long as it remains in the fire. Since the fire constantly infuses the object with heat, heat must also decay in the body. The rate of decay must be proportional to the heat already stored in the body, for that explains the terminal heat—it is the heat at which decay exactly counterbalances the source's heating rate.

Proper application of this to upward moving heavy objects requires a careful consideration of the effect of weight. Motion up starts only once the effective projecting force exceeds the weight of the body. The motion will accelerate as long as the mover keeps increasing the excess over weight of effective moving force at a higher rate than the effective force's rate of decay, and will reach terminal velocity when the effective force decays at the same rate of force infusion by the external mover.

---

[2]Medium resistance could be considered as a growing function of speed that needs to be subtracted from the effective force. This assumption eliminates Aristotle's objection that in a vacuum, the smallest driving force would move the largest body at an infinite speed.

[3]Galilei 1960, pp. 78–79.

Upon removal from the fire, a hot piece of iron begins to cool down gradually. Likewise, upon losing contact with the external source of projecting force, i.e., the mover, the effective force begins to diminish, for it still decays in proportion to its stored quantity. Motion up will continue at a decelerated rate, until the effective force exactly counteracts the weight. But force decay does not stop there. The motion, then, will pass through a single instant of rest, and the body will proceed to fall at an accelerated rate, until all the effective force is spent. At this point terminal down motion obtains, and the body will continue descending at a constant terminal speed that reflects its weight. Weight, then, is not an external force builder. Weight is a tendency inherent to the body—it is constant and characteristic of the body, and if unopposed, it will move the body down at a constant speed proportional to its intensity.

The particular case of bodies let fall from rest is now clear. Up until some moment, they have been kept in place by an external source that infused them or their surrounding medium with effective force equal and opposite to their weight. When let go, they begin descending at an accelerated rate, starting from rest. In other words, this is identical to the case of the upward projected object, from the minute its effective force exactly counterbalanced its inherent weight. And since the infused force decays all the time, the external source must continue to exert itself to keep infusing effective force, even if only to hold the body still.

Simplicius's text says nothing with regard to horizontal motion, but the principles should simply apply with the effect of weight removed. Motion will start the minute the external source begins to infuse effective force into the body or into its surrounding medium. Terminal speed will obtain once the rate of decay   growing proportionately with the growing intensity of effective force—counteracts the rate of force infusion. When the external source ceases its activity, no further infusion of force takes place, but diffusion still continues in proportion to the force already stored. The motion will, therefore, decelerate, until all the force is spent (or falls below the threshold of motion), and rest obtains.

To anyone who finds Aristotle's active medium awkward, the in-body impetus alternative must represent a considerably better-streamlined theory. The medium now only resists motion. Furthermore, the new theory opens up the possibility of admitting the existence of void, and raises the related question of whether in the void the infused force in the body would dissipate. If the force decays only on account of diffusing throughout the surrounding matter, then motion in the void will continue with constant speed; if it decays on its own, being unnatural to the body in the first place, then even in the void projectile motion can only be temporary.

To a loyal Aristotelian, however, these advantages come at an unacceptable price. To begin with, in-body impetus cannot accommodate Aristotle's requirement that a period of rest must intervene between reversals of motion. Without the dynamical interaction between force-carrying medium shells, the falling giant boulder that collides with an upward hurled grain of dust must stop its fall for a brief period to respect the grain's need for a rest period between its upward and downward motion. To accept the theory of in-body impetus therefore requires the rejection of a requirement that Aristotle makes very clearly based on formal considerations. Alternatively, it requires that all matter, no matter how hard, must have a measure of elasticity. More critically, in-body impetus posits self-moving inanimate objects: it endows even a clod of pure elemental earth with self-moving ability. As for the objection that we do not feel the surrounding air pushing us on when executing the long jump, we also do not sense some internal force doing so. We just have a sensation of flying effortlessly on, opposed only by the resistance of the air that both Aristotle and his impetus theory rivals factor in, albeit in different ways.

Appendix A
Do Heavy Objects Become Heavier as they Approach Their Natural Place?

To think that the simple bodies have a different nature the more or less distant they are from their own place is unreasonable—for what is the difference in saying that they are distant to this or that extent? They will differ in proportion to their increase in distance, but their form will remain the same (*On the Heavens*, 276^{b}22-25, Leggatt 1995).

A little further on, Aristotle adds:

A proof that there is not movement to infinity is the fact that earth, by as much as it is closer to the centre, and fire, by as much as it is closer to the upper place, move more quickly. Now if the movement were unlimited, the speed would be unlimited, and if the speed were, so too would the weight or lightness be; for just as if one body were fast by virtue of being lower, another would be as fast by virtue of its weight, so if the increase in weight were unlimited, the increase in speed would be as well (*On the Heavens*, 277^{a}27-32, Leggatt 1995).

Is it reasonable to understand this as suggesting that as a clod of earth falls from up high and approaches the center of heavy things, its heaviness actually increases?[1]

The first point to get out of the way is that even if the heaviness does increase, this cannot explain the everyday phenomenon of accelerated free fall. In *On the Heavens*, II.xiv.298^{a}15 Aristotle estimates the circumference of the earth as 400,000 stadia. With all uncertainties regarding the precise length of a stadium, this estimate is roughly twice the currently accepted radius of about 6,400 km. Aristotle, then, envisioned the earth as a sphere with a radius of roughly 12,000 km. Every-day falls of a few meters to several hundreds of meters represent an insig-

---

[1]That this is so under some circumstances in Newtonian physics is of no consequence – Aristotle's idea of gravity towards the center of the universe is quite different from the Newtonian concept of mutual attraction between masses according to an inverse square law. Furthermore, if we take Newtonian dynamics as the standard, then Galileo's law of free fall is merely a good practical approximation for falls that are very short relative to the distance to the center. As a general natural law, however, Galileo's law is quite false: the acceleration of free fall increases as the object approaches the surface of the earth, and decreases as it falls through the surface toward the center. The object gains weight as it approaches the surface, and loses weight as it continues below the surface towards the center. The mass remains constant, and it is the mass, not the weight, that measures the material content of the body. All of this, of course, is anachronistic and irrelevant to a proper interpretation of either Aristotle or Galileo.

© The Author(s) 2015

I. Yavetz, *Bodies and Media*, SpringerBriefs in History of Science and Technology, DOI 10.1007/978-3-319-21263-0

nificant distance traveled toward the center. Even a fall into a kilometer deep canyon would be accompanied by a mere 1/12000 increase in heaviness. The associated change in heaviness, then, is insignificant, and the clearly observed acceleration of fall could not possibly be accounted for by such negligible increase. Aristotle actually accounts for accelerated natural fall by the intervening medium that always aids natural motion.

Throughout the *Physics* and *On the Heavens*, heaviness and lightness are the thumbprints of the elements. In the case of earth, which is absolutely heavy and has absolutely no lightness to it, the requirement above may be accommodated: no matter how far from the center it is removed, and no matter how much heaviness it loses as a result, it can never acquire lightness, and hence never loses its separate qualitative identity. With regard to water, however, the situation become considerably more murky: it possesses both heaviness and lightness, which means that with sufficient removal from the center, it can always lose enough heaviness and acquire enough lightness to become air, which differs from water in its relative quantities of heaviness and lightness. But this contradicts Aristotle's explicit requirement that removal from the center could not alter the formal essence of the elements.

As it turns out, however, this difficulty need not arise in the first place, considering that the purpose of these statements is to argue against the possibility of an infinite cosmos. The absolute heaviness and lightness of earth and fire, according to Aristotle, express absolute tendencies toward the center and rim respectively: the greater the distance between the center and rim, the greater the difference between heaviness and lightness. Absolute heaviness and lightness, then, increase with the size of the universe, and consequently so will the speed of natural motion up and down for light fire and heavy earth, respectively. This increase in speed is fundamental, and has nothing to do with the acceleration of natural motion that comes in owing to the mediation of the surrounding medium. For a universe of given size, the heaviness of earth as an element is fixed, then. It does not vary with changing distance from the center within this given universe, because earth is earth, fire is fire, water is water, and air is air, regardless of their forced displacement from their natural situations. The size of the cosmos, Aristotle should be understood as saying, cannot be infinite, because in such a cosmos earth will be infinitely heavy and fall infinitely fast, while fire will be infinitely light and float upward infinitely fast. When, in these passages, Aristotle speaks of a body being lower, he is referring to its lower natural place in the order of things, not to its momentary situation relative to that place. There is simply no need to read these passages as suggesting that the fundamental properties heaviness and lightness change value for bodies according to where they happen to be from one moment to the next.

Appendix B
A Threshold of Motion in Time, as well as in Force?

The first part of *Physics* VII.v establishes the basic proportions between moving forces, loads, distance and time, and qualifies them under the condition that motion does not occur as long as the moving force falls below a threshold of motion. The second part of the chapter proceeds to generalize these ideas for forms of change beyond mere locomotion:

> Then does this hold good of alteration and of increase also? Surely it does, for in any given case we have a definite thing that causes increase and a definite thing that suffers increase, and the one causes and the other suffers a certain amount of increase in a certain amount of time. Similarly we have a definite thing that causes alteration and a definite thing that undergoes alteration, and a certain amount, or rather degree, of alteration is completed in a certain amount of time: thus in twice as much time twice as much alteration will be completed and conversely twice as much alteration will occupy twice as much time: and the alteration of half of its object will occupy half as much time and in half as much time half of the object will be altered: or again, in the same amount of time it will be altered twice as much.
>
> And it is the same with qualitive modifications and with growth. For there is something that causes the growth and something that is made to grow, and the process takes so much time, and the growth effected and acquired is so much. So too with the qualitive modification and the quality modified, for a certain 'so much,' as measured by 'more and less,' is modified, and in 'so much' time. Thus if

	A alters B to a degree C in time D,
then	A alters B to a degree 2C in time 2D.
Or	A alters B to a degree $1/2$C in time $1/2$D
and	A alters 2B to a degree C in time 2D
while	A alters $1/2$B to a degree C in time $1/2$D
or	A alters $1/2$B to a degree 2C in time D.

If, on the other hand, there is a force causing alteration or growth, and if

	A alters B to a degree C in time D
and	A alters $1/2$B to a degree C in time $1/2$D
or	A alters B to a degree $1/2$C in time $1/2$D,

© The Author(s) 2015
I. Yavetz, *Bodies and Media*, SpringerBriefs in History of Science
and Technology, DOI 10.1007/978-3-319-21263-0

it does not necessarily follow that ½A will alter B to a degree C in 2D, but it may well happen that ½A will effect no change or growth whatever, just as in the case of the load (*Physics*, VII.v.250^{a}28-250^{b}7, Wicksteed and Cornford 1929).[2]

On the other hand if that which causes alteration or increase causes a certain amount of increase or alteration respectively in a certain amount of time, it does not necessarily follow that half the force will occupy twice the time in altering or increasing the object, or that in twice the time the alteration or increase will be completed by it: it may happen that there will be no alteration or increase at all, the case being the same as with the weight (*Physics*, VII.v.250^{a}28-250^{b}7, Hardie and Gaye 1930).

Is the situation similar in the case of an alteration and an increase? For there is (a) that which causes an increase, (b) that which is being increased (c) so-much time taken, and (d) one thing causing some quantity of increase while another is being increased by that quantity. Similarly there is that which causes an alteration, that which is being altered, a quantity of alteration with respect to more or less, and a quantity of time taken, twice as much [alteration] taking twice the time while in twice the time twice as much [alteration] taking place, and half the alteration taking half the time or in half the time half the alteration taking place, or in equal time twice as much. But if that which causes an alteration or increase does so-much of it in so-much time, it does not follow that half of it will (a) cause it [i.e. the same alteration or increase] in twice the time or will (b) cause half of it in an equal time, but it may happen that it will cause no alteration or no increase at all, as in the case of weights (*Physics*, VII.v.250^{a}28-250^{b}7, Hippocrates G. Apostle 1969).

So then is it the same way also with alteration and increase? For there is something that causes increase and something that is increased, and in so much time, the one causes, and the other undergoes, so much increase. And what causes alteration and what is altered are similar—something *may* be altered so much with respect to more or less, and in so much time, and if in double the time, twice as much, or if twice as much, in double the time, while if half as much, in half the time (or if in half the time, half as much) or in an equal time, twice as much. But if the thing that causes alteration should cause so much in so much time, or the thing that causes increase cause so much in so much time, it is not necessary that either cause half as much in half the time, or if it cause half as much that it be in half the time, but it may perhaps be that it will not cause alteration or increase at all, just as with moving the heavy thing (*Physics*, VII.v.250^{a}28-250^{b}7, Sachs 1995).

So does the same go for alteration and increase? Yes, because there is a certain agent of increase and a certain objet increased, and the one causes increase and the other is increased in a certain amount of time and to a certain extent. Likewise for the agent of alteration and the object of alteration too: there is something which is altered to a certain extent (defined in terms of degree) and in certain amount of time. Twice as much alteration takes twice as much time, and an object twice the size takes twice as much time to alter; an object half the size takes half as much time to alter, and in half the time there will be half as much alteration {of the whole object} and in the whole time an object half the size will alter twice as much. However, the fact that the agent of alteration or of increase causes such-and-such an amount of alteration or increase in such-and-such an amount of time does not make it inevitable that it will alter or increase an object half the

---

[2]The Greek text does not have Cornford's lettered formulae. His translated rendition alters the textual style, but not necessarily its meaning. Furthermore, in the preceding discussion regarding mover, mobile, and locomotion, the Greek text itself uses lettered proportional formulations, so that Cornford's rendition is at least in keeping with the overall flavor of the chapter.

size in half the amount of time,[3] and will cause half as much difference in half the amount of time; no, it may well be that it will cause no alteration or increase at all, which was what we found in the case of weight (*Physics*, VII.v.250^{a}28-250^{b}7, Waterfield 1996).

Is it, then, also thus in alteration and growth? For there is something which causes growth, something which is growing/in so much time, and the one causes so much growth, the other grows so much. And so too what causes alteration and what is altered; something also is altered so much/according to more and less, and in so much time.

In the double time [it will have] a double [effect], and [if it has a] double [effect], [it will do so] in a double time; and half in the half time or, in a half, half, or, [if the power be doubled,] in an equal time, a double [effect]. If, however, what causes alteration or growth causes so much growth or alteration in so much time/it is not also necessary that [the half] cause a half in the half or in the half, a half, but it will alter or will grow nothing, if it chanced, just as also in the case of weight (*Physics*, VII.v. 250^{a}28-250^{b}7, Coughlin 2005).

Beyond obvious and unavoidable stylistic differences, the first three translations of the paragraph (Hardie and Gaye, Cornford and Wicksteed, Hippocrates G. Apostle) diverge markedly from the translations by Sachs and Waterfield (with regard to Coughlin, see below). The first three read the concluding clause in lines 250^{b}4-7 as perfectly analogous to the threshold of locomotion that Aristotle posits two paragraphs earlier. That is, if A displaces (or increases or alters) B by an extent C in time D, it does not necessarily follow that ½A will displace (or increase or alter) B by an extent ½C in time D, (or by an extent C in time 2D) because it may not be able to displace (or increase or alter) it at all, just as the single hauler could not budge the ship that 100 haulers can pull. Sachs and Waterfield by contrast, read no reference to a mover of half the power in the final clause. Instead, they add an interesting new angle to the threshold of motion: If the same movers that cause growth or alteration in a given time operate for half the time only, they will not necessarily cause half the change (alteration or growth as the case may be), because in half the time they may not cause any change at all. This would be a welcome addition, considering that under exposure to sunlight for a given amount of time a plant may grow to a certain degree, while under exposure for half the time it may die altogether rather than grow to half the extent. Likewise, should a physician cut medical treatment short, then rather than healing the patient part way, the treatment may have no healing effect at all. There is a sense of power accumulation over time here that the relationships at the beginning of the chapter do not indicate. The question is, however, how Aristotle's paragraph could be given such divergent readings.

---

[3]Waterfield properly translates "an object half the size takes half as much time to alter," but then directly contradicts it with "...the fact that the agent of alteration or of increase causes such-and-such an amount of alteration or increase in such-and-such an amount of time does not make it inevitable that it will alter or increase an object half the size in half the amount of time". The contradictory meaning Waterfield imported into the second quotation is not in the Greek text, and seems to me mistaken.

A quick look at the manuscript text of lines 250^b4-7 (as Cornford gives it in his explanatory footnote without his emendations) will reveal the cause of the difference:

Εἰ δὲ τὸ ἀλλοιοῦν ἢ αὖξον τὸ τοσόνδε ἐν τῷ τοσῷδε ἢ αὖξει ἢ ἀλλοιοῖ, οὐκ ἀνάγκη καὶ τὸ ἥμισυ ἐν ἡμίσει καὶ ἐν ἡμίσει τὸ ἥμισυ, ἀλλ᾽ οὐδέν, εἰ ἔτυχεν, ἀλλοιώσει ἢ αὐξήσει, ὥσπερ καὶ ἐπὶ τοῦ βάρους

Translated without regard to meaning and proper English expression, the text would read:

If the alterer or increaser, to some extent in some extent either increases or alters, there is no necessity that also the half in half and in half the half, rather nothing, if it chanced, will it alter or increase, just as in the case of the load.

Chapter 5 of *Physics*, Book VII, opens with the following declaration:

Since a mover always moves something, in something, and to some extent (and by in something, I mean in time, and by to some extent, I mean that there is some how much of a distance,… (249^{b}27-28, Sachs).

Aristotle sticks to this formula throughout the chapter, so one may rest reasonably secure with the assumption that whenever "in" appears, (as "in half," in lines 250^b4-7) it always signifies a period of time. Equally common throughout the chapter is the repeated reversal of order between two parameters, e.g. a force may alter an object by the extent D in the extent C, and in the extent C by the extent D. This "iff"-like structure (namely, if A then B and if B then A) is Aristotle's way of securing a single-valued relationship between two parameters. None of this, however, helps in deciding what "the half" in lines 250^b4-7 refers to: is it half the power of the agent, half of the changing object, or half the extent of change?

In an unusually long explanatory footnote that Cornford added to his rendition, he notes that as written, the Greek manuscript does not seem to make reasonable sense, because if καὶ τὸ ἥμισυ indicates half the power, then obviously it couldn't effect the change in half the time. To remain consistent with previous instruction, half the power could effect (1) the whole change in twice the time, (2) half the change in the whole time, or (3) nothing at all, should its strength fall under the threshold of motion. Why then, would Aristotle say that half the power *might* not effect either the full change or half the change in *half the time*, when it is already clear that in half the time it *could not possibly* do either? Coughlin's inserted [the half] in "it is not also necessary that [the half] cause a half in a half…" creates precisely this trivial redundancy, against which Cornford advanced a good argument. On the other hand, Cornford argues, if καὶ τὸ ἥμισυ indicates half the body, then denying that the mover should effect the change in half the time directly violates the rules as given so far. Faced with these difficulties, Cornford apparently took his cue from the concluding words, and reconstructed the paragraph to be analogous to the threshold of motion as clearly stated in the case of displacing loads. In other words, half the power operating on the object may effect no change at all. Hardie and Gaye went this way too. Hippocrates G. Apostle explicitly explains that he took the concluding words as a clue to rendering the text: "There may be some corruption in lines 250^{b}4-7, so we translated by analogy, using lines 250^{a}12-9 {sic}"[4]

As Cornford observed, τὸ ἥμισυ in the phrase οὐκ ἀνάγκη καὶ τὸ ἥμισυ ἐν ἡμίσει καὶ ἐν ἡμίσει τὸ ἥμισυ could not sensibly be understood as referring to half the power, for it is trivially true that half the power could do neither the whole job nor half of it in half the time (ἐν ἡμίσει can only refer to the time in which the job is done, as clearly defined by Aristotle in the opening lines of the chapter). Taking τὸ ἥμισυ to refer to half the body undergoing change is even worse, for then the statement claims that the full power will not change half the body in half the time, directly contradicting the principles formulated so far. Taking τὸ ἥμισυ as referring to half the effect would suggest that the full force operating on the full body will achieve "not necessarily half [the effect] in half [the time] and in [half the time] half [the effect]." Cornford labels this possibility "nonsense," perhaps because it directly contradicts his translation of lines 250^{a}4-5:

> Again if A will move   B over distance   C in time   D
> And A will move   B over distance   ¹/₂C in time   ¹/₂D...

The text says:

Καὶ εἰ ἡ αὐτὴ δύναμις τὸ αὐτὸ ἐν τῳδὶ τῷ χρόνῳ τοσήνδε κινεῖ καὶ τὴν ἡμίσειαν ἐν τῷ ἡμίσει...

Apostle has:

And (3) if the force of A causes B to move over the length S in time T, it also causes B to move over half of S in half the time of of {sic} T,...

Similarly Waterfield:

And if the same power moves the same object just such a distance in just such a time, and half the distance in half the time...

Coughlin also follows suit with:

And if the same power moves the same thing so much in this time and through the half in half...

Sachs, however, seems to have taken τὴν ἡμίσειαν to refer not to half the distance, but to half the body:

"And if the same power moves the same thing this far in this time, and half the thing that far in half the time..." (note, however, that "that far" following "half the thing" is an insertion which is not in the Greek text)

Sachs's translation eliminates any pre-stated condition that if a power moves a body to some distance in some time, then it would move the same body half the

---

[4]Hippocrates G. Apostle, *Aristotle's Physics*, (Bloomington and London: Indiana University Press, 1969), p. 311, note 8.

distance in half the time (and in half the time half the distance). This opens the possibility of understanding the problematic lines 250ᵇ4-7 to suggest that if a mover causes a body to change so much in so much time, it does not necessarily follow that it would cause the body to change half as much in half the time, or in half the time half as much, for it may not produce on the body any change at all if allowed to operate for half the time only. Perhaps this is why Sachs inserted an emphasized *may* in his translation of the following:

καὶ τὸ ἀλλοιοῦν καὶ τὸ ἀλλοιούμενον ὡσαύτως τὶ καὶ ποσὸν κατὰ τὸ μᾶλλον καὶ
ἧττον ἠλλοίωται, καὶ ἐν ποσῷ χρόνῳ, ἐν διπλασίῳ διπλάσιον, καὶ τὸ διπλάσιον
ἐν διπλασίῳ· τὸ δ' ἥμισυ ἐν ἡμίσει χρόνῳ, ἢ ἐν ἡμίσει ἥμισυ ἢ ἐν ἴσῳ διπλάσιον.
(*Physics*, VII.v.250ᵃ31-250ᵇ4)

And what causes alteration and what is altered are similar—something *may* be altered so much with respect to more or less, and in so much time, and if in double the time, twice as much, or if twice as much, in double the time, *while if half as much in half the time (or if in half the time, half as much)*, or in an equal time, twice as much. (Sachs).

Without the qualifying *may*, the underlined phrase would confirm Cornford's "nonsense" for the option of reconstructing the problematic lines 250ᵇ4-7 to state that in half the time, the agent of alteration or change might not cause any alteration or increase at all.

However, in *Physics* VIII.iii.253ᵇ9-10, Aristotle notes "some people" (supposedly followers of Heraclitus), according to whom "... it is not the case that some things are changing, while others are not, but rather that everything is changing all the time; they claim, however, that this fact goes unnoticed by our senses." (Waterfield 1996). Aristotle rejects this claim for the following reasons:

Our argument here resembles the one about the stones being worn away by dripping water or being split by plants; the fact that the dripping water has displaced or removed a certain amount of the stone does not mean that it removed half of that amount of stone in half of that time. No, it is no different from the case of men haling a ship: although so many drops displace so much stone, a proportion of them may not be able to displace that much stone in any amount of time. It is true that the amount of stone which has been moved is divisible into a plurality of parts, but none of them was moved on its own; the point is that they were all moved together. Clearly, then, the fact that the stone removed is infinitely divisible does not necessarily mean that at any given time some part of it is being removed; all we have is that at some time all of it was removed.

The same goes for any kind of alteration as well. The fact that the object undergoing alteration is infinitely divisible does not mean that the alteration is infinitely divisible too; the alteration may well happen all over, as freezing does. Also when something is ill, there has to be a time when it will get better; the change from being ill to being well does not take place instantaneously, but the only possible end-point of the change is health. So to say that alteration goes on and on continually is a drastic way of disputing obvious facts. The point is that alteration has an opposite as an end-point. (*Physics* VIII.iii.253ᵇ14-30, Waterfield 1996).

And there is the argument similar to this one about the wearing down of stones by dripping, or the breaking apart of them by what grows out of them; for it is not the case, if something has pushed out so far or the dripping has taken away so much, that half as much was [added or taken away] beforehand in half the time, but just as with the ship hauling [250a9-b7], so many drips move so much, but a part of them cannot move so much in any amount of time.[5] What is taken away is divisible into many parts, but none of them is moved separately, but all together. It is clear then that it is not necessary for something always to be going away just because a decrease is infinitely divisible, but it goes away at one time as a whole. And similarly too with alteration of any kind whatever: for if the thing altered is infinitely divisible, it is not the case that for this reason the alteration is too, but it often happens all at once, as freezing does. Also, whenever something gets sick, there must come a time in which it gets well, and it is not transformed in the limit of the time [during which it sickens], and must change into health and not into anything else. So to say that alteration goes on continuously is to disagree too much with the obvious. For alteration is into contraries;... (*Physics* VIII.iii.253^{b}14-30, Sachs 1995).

These observations clearly convey the idea that change is restricted by a time threshold, in addition to the power threshold established by the principle that if a given force produces a given effect in a given time, half the force would not necessarily produce half the effect, for it may not produce any effect at all. Sachs and Waterfield reconstruct lines 250^{b}4-7 to convey this idea, and in an appended note, Waterfield actually cites *Physics* VIII.iii.253^{b}14-30 in support of his reconstruction.[6]

Given the vague character of lines 250^{b}4-7, I see both reconstructions as equally valid and equally too specific. To the extent that in the *Physics* Aristotle recorded notes, arguments and observations that he did not structure into a treatise for publication, the problematic lines 250^{b}4-7 need not be regarded as corrupted text. Rather, they may represent a purposefully ambiguous indication of both cases in a single vague and non-specific brief reminder. In a systematic treatise it would be inexcusable to state that an agent of alteration will alter a subject by half the extent in half the time, and then to state that in half the time the agent may achieve no alteration at all without some concluding observation on how the two statements work together. However, as a pedagogical means for emphasizing a particular point in a lecture, it is both legitimate and highly effective for the lecturer first to state a general rule, and then to undermine it with a qualifying statement. In any case,

---

[5]It is noteworthy that Aristotle rejects the idea that a bit of stone wears away with each drop of water, and prefers the idea that a quantity of stone wares away only after so many drops have hit the stone with no wearing (or softening 253^{b}31) effect. He wishes to classify this with the case of a water surface that needs exposure to freezing cold for some finite minimal time below which it does not freeze, and above which the entire surface freezes all at once. The case of the ship haulers is not analogous to the water drops, because the haulers work in parallel, all at the same time, while the drops work in series, one after the other. To be analogous with the water drops, *n* single haulers should individually pull on the boat one after the other with no effect, and only when hauler *n*+1 goes to work would the boat move. In other words, here as in the problematic lines from Book VII. v, one to one analogy does not hold, unless the idea is to emphasize that the case of time is analogous to the case of force in the generalized sense that each case has its respective threshold of motion. Incidentally, Aristotle would have been pleased to find that while water freezes into ice, both it and the ice pretty much maintain a constant temperature of 0°C despite the colder outside temperature. Similarly throughout the boiling process, both the water and the steam it releases maintain a pretty much constant temperature of 100°C, despite exposure to much hotter flames under the pan. Understanding of these phase changes in terms of latent heat begins only with Joseph Black in the 18th century.

[6]Robin Waterfield, *Aristotle Physics*, (Oxford: Oxford University Press, 1996), p. 286, note 250^{b}7.

"translation" is too misleading a term for these lines. Literally translated, they do not convey a single valued message, and suggestions as to what Aristotle wished to convey by them can be put forward only as speculative reconstructions. As such, the two senses in the six renditions above are equally plausible, and not necessarily mutually exclusive.

Appendix C
A Mathematical Formulation of Aristotle's Theory of Forced Horizontal Motion

For forced motion transverse to the direction of natural motion, Aristotle requires that the momentary speed, V, be directly as the effective moving force of the medium, E, and inversely as the weight, W, of the mobile, and the resistance of the medium. Allowing the coefficient of resistance, ρ, to serve as the constant of proportionality, this becomes:

$$E = \rho W V. \tag{C.1}$$

To reflect the accumulation and decay of medium force, set its rate of change to be directly as the difference between F, the force of the original mover, and E:

$$\frac{dE}{dt} = \lambda(F - E). \tag{C.2}$$

Combining this with the basic relationship between effective force and speed:

$$\rho W \frac{dV}{dt} = \lambda(F - \rho W V),$$

Or:

$$\frac{dV}{dt} + \lambda V = \frac{\lambda F}{\rho W}. \tag{C.3}$$

Equation (C.3) lumps all the features of levels 1, 2, and 3 into a single equation of motion. The solution for $V(0) = 0$ and a constant external mover F, is:

$$V(t) = \frac{F}{\rho W}\left(1 - e^{-\lambda t}\right). \tag{C.4}$$

The speed approaches exponentially a terminal value of $F/\rho W$. At all times the strict proportionality of levels 1 and 2, expressed in Eq. (C.1) above, is preserved.

© The Author(s) 2015
I. Yavetz, *Bodies and Media*, SpringerBriefs in History of Science and Technology, DOI 10.1007/978-3-319-21263-0

For two different loads moved by two different original movers in the same medium, we get:

$$\frac{V_1}{V_2} = \frac{F_1}{F_2} \cdot \frac{W_2}{W_1},$$

So the proportions of *Physics* VII.5 hold for the external mover as well as for the effective moving force.

For the same mobile moved by the same original mover in two different media, we get:

$$\frac{V_1}{V_2} = \frac{\rho_2}{\rho_1} \cdot \frac{1 - e^{-\lambda_1 t}}{1 - e^{-\lambda_2 t}},$$

So under the action of the same original mover, as opposed to the same effective force, the linear inverse proportionality between speeds and medium resistance holds for the terminal speeds only. This is because λ is medium-specific, so the same original mover gives rise to different effective forces in different media. For a given effective force and a given load, the inverse proportionality of speed and resistance applies without exception to variable as well as uniform speeds.

Three features of this theory may be observed at this stage:

1. While effective force is always proportional to speed, the ever-present medium adds an acceleration stage to any motion induced by the application of an original mover to a previously stationary mobile. There is nothing contradictory in this, because the appearance of acceleration is not fundamental to the relationship between force and speed. Rather, it reflects the dynamics of motion in the medium, and a medium is always present. It also becomes clear that assigning to the medium both resistance and support of motion involves no contradiction: force buildup and decay in the medium is independent of the medium's resistance, and they produce perfectly coherent, unambiguous results.

2. The general equation of motion (C.3) re-asserts the difficulties that Aristotle found with the idea of vacuum. Where no medium exists, both ρ and λ, the resistance and coefficient of decay, are 0. Their ratio differs from medium to medium, and no limiting process can reduce to a uniquely defined vacuum ratio. Different media may rarefy into different limiting ratios of ρ and λ, and no conclusion is therefore possible as to the vacuum, which simply means no resistance, no holding capacity, and hence no ratio between two non-existent things.

3. Consider two different loads, brought to the same horizontal speed, and then disconnected from their original movers to move on their own. The right side of Eq. (C.3) must be set to 0, and the theory predicts that the two bodies will lose speed at the same rate. Indeed, it requires no recourse to the differential equation of motion (C.3) to arrive at this conclusion, and Eq. (C.1) that directly reflects Aristotle's proportions in levels 1 and 2 suffices: to move two different loads at

the same speed, the medium layers surrounding the moving loads must apply effective forces that relate as the loads. Left to its own devices after losing contact with the original movers, the ratio of force to load is therefore the same for the two bodies, implying that their speeds must be equal at all time in accordance with the rules of level 1 (*Physics* VII.v). A carefully set experiment will violate this expectation with two polished spheres of equal size and different weight, that start out rolling on a smooth horizontal floor with equal speeds in the same medium. Observation will invariably show that the speed of the heavier one will decay more slowly than the speed of the lighter one.

Appendix D
A Mathematical Formulation of Aristotle's Theory of Natural and Forced Vertical Motion

The essentially new feature of natural fall is that heaviness is not a type of original moving force that pushes a stone downward, nor could it possibly represent a load against its own natural downward tendency. The equation of decay, then, is common to all cases of motion, horizontal as well as vertical:

$$\frac{dE}{dt} = \lambda(F - E). \tag{D.1}$$

With respect to anything that stands in the heavy mobile's way down, the weight does act as an original mover. In the medium, therefore, the heaviness appears as a fixed downward tendency that must be added to the effective force built into it by any external, vertically acting original mover. To reflect this, we associate medium force, heaviness, speed, and the medium's resistance to being cleaved as:

$$E - W = \rho V. \tag{D.2}$$

As before, ρ and λ are respectively the coefficients of resistance and force decay in the medium. In the absence of any external agency, Eq. (D.2) ensures that at all times the speed (variable or fixed) is proportional to the weight, in accordance with Aristotle's explicit requirement.

Solving for E in (D.2) and substituting this into (D.1) yields:

$$\frac{dV}{dt} + \lambda V = \frac{\lambda}{\rho}(F - W). \tag{D.3}$$

This has some remarkable consequences. If $F = W$, and the initial speed is 0, the mobile will remain at rest. If, however, the initial speed was a finite V_0, either directly down or directly up, the mobile's speed will decay to 0 at the same rate and irrespective of the resistance, as long as F is present.

© The Author(s) 2015
I. Yavetz, *Bodies and Media*, SpringerBriefs in History of Science and Technology, DOI 10.1007/978-3-319-21263-0

If the mobile was at rest, and F was suddenly taken away at $t = 0$, the equation of motion becomes:

$$\frac{dV}{dt} + \lambda V = -\frac{\lambda}{\rho} W. \tag{D.4}$$

Going beyond Aristotle to Archimedes's law, the effective weight in a material medium is the difference between the mobile's absolute weight, W_b, and the weight, W_m, of a volume of medium equal to the volume of the mobile, so Eq. (D.4) becomes:

$$\frac{dV}{dt} + \lambda V = -\frac{\lambda}{\rho} W_b \left(1 - \frac{W_m}{W_b}\right). \tag{D.5}$$

The solution of Eq. (D.5) for $V(0) = 0$ is:

$$V(t) = \frac{W_b}{\rho} \left(1 - \frac{W_m}{W_b}\right) \left(e^{-\lambda t} - 1\right) = \frac{W}{\rho} \left(e^{-\lambda t} - 1\right). \tag{D.6}$$

From Eq. (D.6) it follows that for two bodies of different weights and identical forms falling together from rest in any given medium, the proportion $V_1 : V_2 :: W_1 : W_2$ holds at all times, whether the speeds are accelerating or constant. This is what Aristotle requires in his argument against a weightless body falling naturally, which corroborates the appropriateness of the mathematical rendition given here to his theory of natural motion.

Equation (D.6) predicts that were two bodies of the same form, one weighing ten times more than the other, dropped from a tower, then by the time that the heavy one would reach ground the light one would cover only a tenth of the tower's height. Galileo turned this into a crucial argument against Aristotle's theory, claiming that in fact, the heavy body would beat the light one by a mere finger. However, Galileo's argument turns out to be much too hasty, as the following analysis demonstrates.

In *Two New Sciences*, in an argument leading to the constancy in vacuum of natural acceleration for all bodies, Galileo observed that in any heavy, resisting medium, just at the beginning of fall when the speed is practically negligible, all mobiles fall at an acceleration reduced from a general constant by the ratio between the specific weight of the medium and the specific weight of the mobile:

$$\frac{dV}{dt} = -g \left(1 - \frac{W_m}{W_b}\right). \tag{D.7}$$

Equation (D.5) for V = 0 reduces to:

$$\frac{dV}{dt} = -\frac{\lambda}{\rho} W_b \left(1 - \frac{W_m}{W_b}\right),$$

and so from this together with Eq. (D.7) it follows that:

$$\frac{\lambda}{\rho} = \frac{g}{W_b}. \tag{D.8}$$

Since all four quantities here are constants, this relationship, extracted immediately at the onset of motion, must remain invariable at all time. We already noted that the resistance of the medium is not an independent characteristic of the medium, but depends on the geometrical form of the mobile. Now consider an Aristotelian physicist who admits Eq. (D.7) as an experimental discovery by Galileo. To this Aristotelian physicist Galileo's discovery reveals a further restriction in the form of strict proportionality between the ratio of the medium's coefficent of force decay to its resistance and the ratio of the constant of acceleration to a body's weight. So far, then, Galileo's observation reveals a previously unsuspected connection between the relevant properties of the medium, the absolute weight of the mobile, and the maximum acceleration that can be observed at the beginning of natural fall. This would constitute a striking new scientific discovery, but it also gives the lie to Galileo's attempt to turn it into a crucial critique of Aristotelian dynamics, because:

Substituting Eq. (D.8) into (D.5) above yields:

$$\frac{dV}{dt} + \rho \frac{g}{W_b} V = -g \left(1 - \frac{W_m}{W_b}\right),$$

and after multiplying through by W_b/g:

$$\frac{W_b}{g} \frac{dV}{dt} + \rho V + (W_b - W_m) - 0. \tag{D.9}$$

Equation (D.9) is mathematically identical to Newton's equation of motion for a mobile of weight W_b falling freely under its own weight in a medium weighing W_m in a volume equal to the mobile's, and in which resistance to motion grows linearly with speed. For an object with specific weight much greater than the medium's this becomes approximately:

$$\frac{W_b}{g} \frac{dV}{dt} + \rho V + W_b = 0,$$

so for a short time after onset, while V is still small, the accelerations of two mobiles of different weights, both much greater than the medium's, are almost indistinguishable. Even if dropped from a high tower, as long as their speeds remain much less than their terminal speeds, they will reach bottom almost at the same time. This, not by a new and revolutionary seventeenth century theory of motion, but by Aristotle's dynamics, worked into consistent mathematical terms, and quite easily incorporating Galileo's law of free fall. Galileo's attempt to turn the famous Tower of Piza experiment into a refutation of Aristotle merely reflects a failure to methodically work out Aristotle's theory of motion from the hints dispersed throughout the *Physics* and *On the Heavens*.

Now consider again the Aristotelian theory of horizontal motion, and its problematic prediction that horizontally moving projectiles lose speed at the same rate regardless of weight. For an object that moves horizontally under no external mover, starting out at speed V_0, Eq. (C.3) of Appendix C requires:

$$V(t) = V_0 e^{-\lambda t},$$

Where V_0 and λ are respectively the initial speed and the coefficient of force decay in the medium. Now express λ in terms of the new proportion revealed by incorporating Galileo's discovery that free fall has the same acceleration regardless of weight:

$$V(t) = V_0 e^{-\rho \frac{g}{W_b} t},$$

And it turns out that other things being equal, the speed of a horizontally rolling sphere will decay more slowly the heavier it is, in conformity with experience. Not only can Aristotle's theory accommodate Galileo's discovery, but the accommodation actually saves it from an incompatibility with experience in the case of horizontal projectile motion.

The physics of vertical motion may be set up differently, with no effect on the resulting equation of motion. In the alternative version, the external mover and the heaviness compete as generators of effective force in the medium, and hence the heaviness, W, figures in the equation of effective force decay as follows:

$$\frac{dE}{dt} = \lambda(F - W - E), \tag{D.10}$$

Speed, on the other hand, is now related only to the net effective force via the resistance:

$$E = \rho V. \tag{D.11}$$

The resulting equation of motion will be identical to (3) above, but its physical explanation will now be different. The external force, F, either works with or competes against the heaviness, W, to build effective force into the surrounding medium. Therefore if initially there is no moving force in the medium ($E = 0$), then by Eq. (D.11) at that initial time there will also be no motion. If $F = W$, then by

Eq. (D.10) no moving force will accumulate in the medium and the state of rest will persist. Of course, F will continue to strain to build force into the medium to counteract W's activity to build medium force in the opposite direction, which explains why we get tired just trying to hold a heavy object up without moving it.

Now consider a situation where following a state of rest, with $F = W$, F suddenly increases. This will gradually build moving force into the medium, accompanied by accelerating upward motion. Once E attains equality to the difference between F and W, there will be no further change in E (because the sum of factors in parentheses in Eq. (D.10) will be zero), and the upward motion will continue at a constant speed. At constant upward speed, assume that F is suddenly removed—as in the case of contact breaking between a thrower's hand and the thrown object. From this point, upward motion will continue at a decelerating rate, as E diminishes. Rest will be passed through at the instant that $E = 0$, but this cannot last for any length of time, because the unopposed weight will build downward moving effective force with accelerating downward motion. Terminal velocity ensues once $E = -W$, which will reduce the effective force building rate to zero in the absence of any external source of force.

In the previous formulation (Eqs. (D.1) and (D.2)), momentary rest ensues once the effective force exactly counteracts the fixed effective downward tendency called heaviness. The heavy object reaches its downward terminal speed once no more effective moving force remains in the medium, and the fall now reflects the natural heaviness, mitigated only by the medium's resistance to being cleaved. In the second formulation, the momentary rest reflects no effective moving force in the medium, while terminal speed reflects moving force in the medium that is equal to the heaviness, now understood not as an effective force, but as an effective force builder. In the second formulation then, natural downward motion reflects heaviness indirectly, via the effective moving force that heaviness builds into the medium. In the first formulation, natural downward motion reflects heaviness directly as an effective mover in and of itself.

Against the first formulation stands Aristotle's explicit statement that the medium always operates as intermediary in both forced and natural motion. But then, Aristotle immediately continues to say that the medium is solely responsible for (prolonged) forced motion, and aids natural motion. The last clause can be understood as referring to cases when natural fall is accompanied by an additional downward push (like throwing a thing downward, as opposed to merely releasing it). In this case, medium action will aid the natural fall by prolonging the forced addition. The first formulation, according to Eqs. (D.1) and (D.2), also requires that while heaviness exerts cleaving force against the medium, it does not at the same time induce moving force into it, which seems awkward. These, however, are not strong enough objections, so both formulations work well as possible interpretations of Aristotle's text, and in the absence of any clear statement by him as to the preferability of one over the other, any preference seems ill-advised. Both should be kept in mind as equally plausible.

The graphs in Fig. D.1 represent three 45° throws in media with different ratios of force decay (λ) to resistance (ρ), using Eq. (D.4) of Appendix D for the vertical motion, and Eq. (C.3) of Appendix C for the horizontal motion. The plots depict the trajectory after the projectile has left the thrower's hand. The initial velocity is the same in all three cases, but the time and distance scales are different and the graphs merely intend to emphasize the difference that the ratio $\lambda{:}\rho$ makes for the shape of the trajectory.

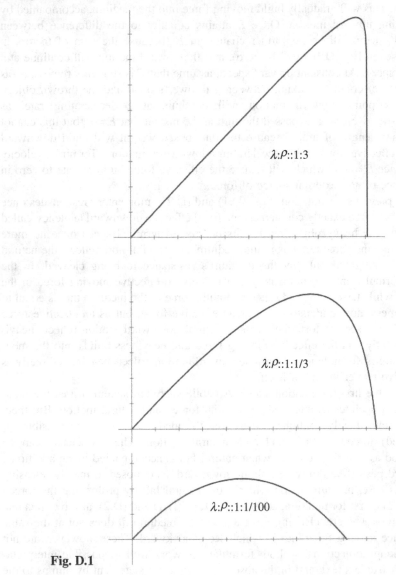

$\lambda{:}\rho{::}1{:}3$

$\lambda{:}\rho{::}1{:}1/3$

$\lambda{:}\rho{::}1{:}1/100$

Fig. D.1

Appendix E
A Mathematical Formulation
of Hipparchus's Theory of Vertical Motion

Hipparchus retains Aristotle's basic connection between effective moving force, E, and speed of motion, V:

$$E = A \cdot V, \tag{E.1}$$

where A is a proportionality constant that is case specific, and takes into account the shape of the body, and any other property of the body and the environment that implicitly affects the motion without being given explicit expression elsewhere.

Coming now to Hipparchus's principle, two properties must be given explicit consideration. (1) Mobiles of equal shapes but different materials moving through the same medium slow down at different rates (compare the flights a ball carved out of cork to an equally sized iron ball). Clearly different materials differ in their capacity, C, to retain moving force, so this is a mobile-specific parameter. (2) Even casual observation of the same body moving in air, water, and honey easily discloses that the same body moving in different media loses speed at different rates. Now, the passage wherein Simplicius describes Hipparchus's theory says nothing about the medium's resistance, and there is actually no need to mention it, for different media may be taken to differ in their ability to dissipate moving force. The specific dissipation parameter, D, then distinguishes between different media. Taking a cue form Galileo's analogy to cooling, and noting that cooling rate is quicker at high temperatures and slower at low temperatures, the following provides the simplest expression of the implied proportion between decay rate and force, f:

$$\frac{df}{dt} = -\frac{D}{C}f. \tag{E.2}$$

Now, however, the role of an external source of motion must figure in, just as a heating fire must account for the heating of a body. This calls for an expression for the rate of change of the internal force, which could either grow or diminish, according to the situation. Dissipation must be proportional to the force already

© The Author(s) 2015
I. Yavetz, *Bodies and Media*, SpringerBriefs in History of Science
and Technology, DOI 10.1007/978-3-319-21263-0

stored in the body; accumulation must be proportional to F, the intensity of the external source. The body's force holding capacity is an internal property that does not distinguish between incoming and outgoing force—it is a change-resisting faculty, and to the extent that it resists losing what it has, it resists adding to it. However, a decision must be made with regard to the diffusion rate. Dissipation has been ascribed to the medium, and so there should be none in a vacuum, where nothing could conduct the force out, so to speak. Can, however, any force diffuse into the body in a vacuum, even when in contact with some external source? If so, there must be two diffusion parameters, D_i and D_o, for incoming and outgoing force, respectively, and then:

$$\frac{df}{dt} = \frac{D_i}{C}F - \frac{D_o}{C}f.$$

(E.3)

However, it may also be assumed that where no diffusion out exists, neither does diffusion in. It is always via the surrounding medium that fire heats and external power sources infuse a body with moving force. In this case:

$$\frac{df}{dt} = \frac{D}{C}(F - f).$$

(E.4)

Because of the way it relates back to Aristotle's ideas, and forward to Galileo's, this option will be developed here. (There is no pretense to suggest that this is what Hipparchus had in mind, because in any case he probably did not have in mind anything like these mathematical formulations).

Finally, it would seem implausible to assume that Hipparchus did not know Archimedes's way of estimating the effective weight, W, of a body immersed in some material medium:

$$W = W_b - W_m,$$

(E.5)

where W_b is the body's weight, while W_m is the weight of the medium in a volume equal to the body's. The effective force on the body is the difference between its internally stored force, f, and its effective weight:

$$E = f - w_b\left(1 - \frac{W_m}{W_b}\right) \ \Rightarrow\ f = E + W_b\left(1 - \frac{W_m}{W_b}\right).$$

(E.6)

Using this in Hipparchus's decay principle with a single diffusion coefficient (Eq. E.4):

$$\frac{dE}{dt} = \frac{D}{C}\left[F - E - W_b\left(1 - \frac{W_m}{W_b}\right)\right].$$

(E.7)

With Aristotle's basic connection between effective force and speed (Eq. E.1):

$$A\frac{dV}{dt} = \frac{D}{C}\left[F - AV - W_b\left(1 - \frac{W_m}{W_b}\right)\right],$$ (E.8)

and, finally:

$$\frac{dV}{dt} + \frac{D}{C}V + \frac{DW_b}{AC}\left(1 - \frac{W_m}{W_b}\right) = \frac{D}{AC}F.$$ (E.9)

Or, if the distinction between D_i and D_o is maintained (Eq. E.3):

$$\frac{dV}{dt} + \frac{D_o}{C}V + \frac{D_oW_b}{AC}\left(1 - \frac{W_m}{W_b}\right) = \frac{D_i}{AC}F$$ (E.10)

The last two equations replace Aristotle's Eq. (C.3) in Appendix C.

Simplicius's text says nothing about horizontal motion, so it may be assumed that the above equation should be used, but with the body's weighted downward trend exerting no influence, and instead figuring into Aristotle's basic relationship (1) as a load inversely proportional to the speed, that is:

$$E = AW_b \cdot V,$$ (E.11)

and then:

$$\frac{dV}{dt} + \frac{D}{C}V = \frac{D}{AW_bC}F.$$ (E.12)

Or, with the distinction between D_i and D_o:

$$\frac{dV}{dt} + \frac{D_o}{C}V = \frac{D_i}{AW_bC}F.$$ (E.13)

Should a vacuum be allowed (Simplicius's text says nothing about this with regard to Hipparchus), then $D = 0$ for the single diffusion coefficient suggests that no acceleration is possible in a vacuum. Bodies at rest will remain so forever, bodies in motion will continue their motion forever at a constant speed. In the split parameter case, $D_o = 0$ for the medium's diffusion coefficient suggests that constant speed is the vacuum case when no external source of motion applies. When it does apply, the motion will have a constant acceleration in proportion to the external source's intensity. In either case, the vacuum completely eradicates the influence of heaviness.

Now, however, factor in Galileo's observation that at the onset of free fall, when the speed is very low, all bodies accelerate according to (the (–) sign indicating downward trend):

$$\frac{dV}{dt} = -g\left(1 - \frac{W_m}{W_b}\right). \tag{E.14}$$

Using this value for the acceleration in Eq. (E.9) in the case of release from rest (external force $F = 0$) restricted to the initial stage of motion where V is negligible:

$$g\left(1 - \frac{W_m}{W_b}\right) = \frac{DW_b}{AC}\left(1 - \frac{W_m}{W_b}\right),$$
$$\Rightarrow g = \frac{DW_b}{AC}. \tag{E.15}$$

Since D, W_b, A, and C are all constants independent of speed, Eq. (E.15) must hold for all speeds, and since W_b and C are characteristics of the body regardless of either speed or environment, this reveals a dependence between the two environmental coefficients, D and A (being the diffusion coefficient, and the general Aristotelian proportionality constant that encompasses both properties of the body and the environment):

$$D = \frac{gC}{W_b}A. \tag{E.16}$$

That is to say, it now turns out that in approaching vacuum, D and A diminish at the same rate so as to keep their ratio a constant. Putting this back into the general equation of motion:

$$\frac{dV}{dt} + \frac{g}{W_b}AV + g\left(1 - \frac{W_m}{W_b}\right) = \frac{g}{W_b}F, \tag{E.17}$$

or:

$$\frac{W_b}{g}\frac{dV}{dt} + AV + (W_b - W_m) = F, \tag{E.18}$$

which, once again, is mathematically equivalent to the Newtonian equation of motion in a medium that offers resistance proportional to the speed. In a vacuum $W_m = 0$, and there is no diffusion into a receiving medium, so $D = 0$, which now, through Eq. (E.16) requires also $A = 0$. Using all of this in Eq. (E.18) reveals that in a vacuum, all bodies regardless of weight fall naturally with the same constant acceleration, g. Note also that inclusion of Galileo's law of free fall argues strongly against splitting the infusion and diffusion coefficients, because in that case, the condition in Eq. (E.14) applies to Eq. (E.10), showing that Eq. (E.15) holds for D_o only, and hence:

$$\frac{W_b}{g}\frac{dV}{dt} + AV + (W_b - W_m) = \frac{D_i}{D_o}F, \qquad (E.19)$$

so that if a vacuum is allowed, it turns out that any finite external force, no matter how small, would generate infinite acceleration in all bodies of finite weight, no matter how large.

Appendix F
Alternative Translations of the Quotations Used in the Main Text

Physics VII.v.249^{b}27-250^{a}9

That which is causing motion is always moving something in something and up to somewhere. (By the something 'in which' I mean time, and by the 'up to somewhere' the measure of the distance traversed; for if a thing is now causing motion, it has already caused motion before now, so that there is always a distance that has been covered and a time that has been taken.) If, then, A is the moving agent, B the mobile, C the distance traversed and D the time taken, then

A will move ½B over the distance 2C in time D, and
A will move ½B over the distance C in time ½D;
For so the proportion will be observed.
Again if
A will move B over distance C in time D and
A will move B over distance ½C in time ½D, then
E (= ½A) will move F (= ½B) over distance C in time D;

For the relation of the force E (½A) to the load F (½B) in the last proposition is the same as the relation of the force A to the load B in the first, and accordingly the same distance (C) will be covered in the same time (D) (Wicksteed and Cornford 1929).

Now since wherever there is a movent, its motion always acts upon something, is always in something, and always extends to something (by 'is always in something' I mean that it occupies a time: and by 'extends to something' I mean that it involves the traversing of a certain amount of distance: for at any moment when a thing is causing motion, it also has caused motion, so that there must always be a certain amount of distance that has been traversed and a certain amount of time that has been occupied). then, A the movement [*sic*] have moved B a distance G in a time D, then in the same time the same force A will move 1/2B twice the distance G, and in 1/2D it will move 1/2B the whole distance for G: thus the rules of proportion will be observed. Again if a given force move a given weight a certain distance in a certain time and half the distance in half the time, half the motive power will move half the weight the same distance in the same time. Let E represent half the motive power A and Z half the weight B: then the ratio between the motive power and the weight in the one case is similar and proportionate to the ratio in the other, so that each force will cause the same distance to be traversed in the same time (Hardie and Gaye 1930).

Now since a mover always moves something and is in something, and extends to something (by 'is in something' I mean that it occupies a time; and by 'extends to something' I mean that it involves a certain amount of distance—for at any moment when a thing is causing motion, it also has caused motion, so that there must always be a certain amount of distance that has been traversed and a certain amount of time that has been occupied). If, then, A is the mover, B the moved, C the distance moved, and D the time, then in the same time the same force A will move 1/2B twice the distance C, and in 1/2D it

© The Author(s) 2015
I. Yavetz, *Bodies and Media*, SpringerBriefs in History of Science and Technology, DOI 10.1007/978-3-319-21263-0

will move 1/2B the whole distance C; for thus the rules of proportion will be observed. Again if a given force moves a given object a certain distance in a certain time and half the distance in half the time, half the motive power will move half the object the same distance in the same time. Let E represent half the motive power A and F half B: then they are similarly related, and the motive power is proportioned to the weight, so that each force will cause the same distance to be traversed in the same time (Barnes 1984).

An agent of movement always moves something, does so in something, and does so to some extent. By 'in something' I mean 'in some time', and by 'to some extent' I mean 'over a certain amount of distance', because at one and the same time it is causing movement and has caused movement, so that there will always be a certain distance which has been moved and a certain period of time that has been taken. So let A be the agent of movement, B the object moved, C the distance it has moved, and D the amount of time it has taken. The ratios will be preserved, then, if in an equal amount of time an equal power A moves half B double the distance C, and moves half B over C in half D. And if the same power moves the same object just such a distance in just such a time, and half the distance in half the time, then half the power will take an equal amount of time to move an object half the weight over an equal distance. For example, let E be a power which is half A and F be an object which is half the weight of B. The two examples are similar—the power and the weight are in the same ratio in both cases—and so the two powers will move their respective objects over an equal distance in an equal time (Waterfield 1996).

Since, however, the mover always moves something and in something and up to something (I say, however, 'in something' because it is in time, 'up to something' because the length is some amount, for the mobile always is moving and has moved at the same time, whence, there will be some amount which the mobile is moved, and in so much time), if, then, A is the mover, B the moved, the length C how much it has been moved, in so much, the time D, then, in an equal time a power equal to that of A will move the half of B double C [it will move half of B through] the length C, however, in that half of D. For thus will there be proportion.

And if the same power moves the same thing so much in this time and through the half in half, the half strength will move the half in an equal time through an equal length. For example, let E be half the power of a and Z the half of B. The strength, then, is disposed similarly and according to proportion in relation to the weight; whence, it will move [Z through] an equal length in an equal time (Coughlin 2005).

Physics, IV.viii.215^{a}29-215^{b}12

Now the medium reduces velocity all the more if it is in itself moving in the opposite direction, but it also reduces it, though in a lesser degree, if it is quiescent; and this impedium of motion is proportional to the resistance the medium offers to cleavage, which is to say its density. Thus, if one medium is easier to cleave than another, the time taken in travelling a given distance through it will be proportionately less. *E.g.*, if the media are water and air, the ratio of air's cleavableness and unsubstantiality to that of water will give the ratio of the velocity of the passage through air to that though water.

Velocity in air : velocity in water = Cleavability of air : cleavability of water.

So, if air is wice as easy to cleave as water, the passage through air will be twice as swift, and the time taken in covering a given distance half as long in air as in water. According to this universal principle, then, the velocity will in every case be greater in proportion to the unsubstantiality and diminished power of impeding and easier cleavability of the medium (Wicksteed and Cornford 1929).

Now the medium causes a difference because it impedes the moving thing, most of all if it is moving in the opposite direction, but in a secondary degree even if it is at rest; and especially a medium that is not easily divided, i.e. a medium that is somewhat dense. A, then, will move through B in time G, and through D, which is thinner, in time E (if the length of B is equal to D), in proportion to the density of the hindering body. For let B be water and D air; then by so much as air is thinner and more incorporeal than water, A will move through D faster than through B. Let the speed have the same ratio to the speed, then, that air has to water. Then if air is twice as thin, the body will traverse B in twice the time that it does D, and the time G will be twice the time E. And always, by so much as the medium is more incorporeal and less resistant and more easily divided, the faster will be the movement (Hardie and Gaye 1930). [7]

The medium through which the body travels is a cause by the fact that it obstructs that body, most of all if it [the medium] is travelling in the opposite direction, but even if it is *resting*; and it does so more if it is not easily divisible, and such is a more viscous medium. [8]

The body A, then, will travel through medium B in time C and through medium D (which is less viscous) in time E, these [C and E] being proportional to the obstructing medium, if the length of B and D are equal. For let B be water and D be air. Then the extent to which air is less viscous or less corporeal than water is proportional to the extent to which A travels faster through D than through B. Let the two speeds have the same ratio as that by which air differs from water. Then if air is half as viscous as water, A will travel though B in twice the time as it will through D, and C will be twice as long as E. And always, the more incorporeal or less obstructive or more easily divisible is the medium, the faster will the body travel through it (Hippocrates G. Apostle 1969).

Now the medium through which it moves is responsible because it resists the thing, especially when it itself is moving in the opposite direction, but even when it is stationary. What is not easily divisible resists more, i.e. what is of thicker texture. So [body] A will move through [medium] B in time C, and through D, which is of finer texture, in time E, in proportion to the resisting body, (assuming that the length of B is equal to that of D). For example, let B be water and D be air. A will move faster through D than through B by the

---

[7]The parenthetical remark "if the length of B is equal to D" is in the Greek text. "In proportion to the density of the hindering body" adds a reference to "density" which is not in the Greek (Barnes's 1984 edition of Aristotle's complete works retains Hardie and Gaye's translation for this paragraph). The text actually says κατὰ τὴν ἀναλογίαν τοῦ ἐμποδίζοντος σόματος, "in proportion to the hindering body," namely, proportional to the hindrance of the medium, which is quite sensible without the added reference to density. To reflect the structure of the text, reference to the density, or viscosity, and preferably to the thickness of the medium should appear before the proportional relationship, not within it (as in Apostle, Hussey, Sachs, Waterfield, and Coughlin). While Hardie and Gaye, Wicksteed and Cornford, and Waterfield use "density" throughout chapters 8 and 9, the Greek text makes a very clear distinction between λεπτός and παχύς, thin and thick, which serve only chapter 8, as opposed to μανός and πυκνός, rare and dense, which serve throughout chapter 9. The distinction is clearly noted and carefully observed by Edward Hussey, (1983) who uses "fine texture" and "thick texture" for the first pair, rare and dense for the second. Sachs (1995) uses thin and thick and rare and dense, respectively, while Coughlin (2005) prefers subtle and thick for the first pair, rare and dense for the second). The first pair is therefore clearly related to a dynamic concept—the resistance to motion offered by a material medium, while the second pair serves the discussion of the magnitude of space associated with a given portion of matter.

[8]It is definitely a point in Aristotle's favor that he did not fall into the error of assuming that the resistance to motion in a medium is a function of its density alone. However, it still requires a very significant conceptual step from this to the specific dynamical roles that density and viscosity play in classical fluid dynamics. Apostle's use of the term "viscous" in this context may, therefore, be a little too charitable. See also Apostle's translation of Aristotle's *Physics*, (1969), note 16, p. 255.

amount by which air is of finer texture and less corporeal than water. So let the speeds have the same proportion one to another as that in which air differs from water, so that, if air is twice as fine in texture, it will traverse B in twice as much time as it takes to traverse D, and the time C will be double the time E. And so it will always be that the body will move faster by the amount by which the medium through which it moves is less corporeal and less resistant and more easily divisible (Hussey 1983).

Now, the medium through which the object is moving makes a difference by impeding the object. It does so especially if it is moving in the opposite direction but also if it is still. The resistance is greater if the medium is not easy to divide, which is to say, if the medium is denser. An object A will move through B in time C, but through the less dense medium D in time E (assuming that B and D are equal in extent), and the times will be proportionate to the resistance exerted by the impeding body. Let B be water and D air; in proportion as air is less dense and less material than water, A will move this much faster through D than it does though B. Let the speeds have the same ratio to each other, then, as the density of air does to the density of water, so that if air is twice as refined, A will take twice as long to traverse B as it does to traverse D, or in other words the time C will be double the time E. And it will always be the same case that in proportion as the medium is less material, less impeding, and easier to divide, A will move that much faster (Waterfield 1996).

That through which the mobile is borne, then, is a cause because it impedes, most when being borne against [the mobile], but also if remaining {resting}; and what is not easily divisible [impedes] more, and such is the thicker [stuff]. A, then, will be borne through B in time C, but through D, being more subtle, in E, according to the proportion of the impeding body, if the length of B is equal to D. For let B be water and D air; then as much as air is more subtle and more unbodily than water, so much faster will A be borne through D than through B. Let the speed stand to the speed, then, in the same ratio as the air to the water. Whence, if it is twice as subtle, it will go through B in twice the time it goes through D, and time C will be double E. And always, then, as much as that though which it is borne is more unbodily and less impending and more easily divided, [so much] faster will it be borne (Coughlin 2005).

Physics VIII.x.266^{b}28-267^{a}21

But before discussing rotating bodies it will be well to examine a certain question concerning bodies in locomotion. If everything that is in motion is being moved by something, how comes it that certain things, missiles for example, that are not self-moving nevertheless continue their motion without a break when no longer in contact with the agent that gave them motion? Even if that agent at the same time that he puts the missile in motion also sets something else (say air) in motion, which something when itself in motion has power to move other things, still when the prime agent has cease to be in contact with this secondary agent and has therefore ceased to be moving it, it must be just as impossible to it as for the missile to be in motion: missile and secondary agent must all be in motion simultaneously, and must have ceased to be in motion the instant the prime mover ceased to move them; and this holds good even if the prime agent is like the magnet, which has power to confer upon the iron bar it moves the power of moving another iron bar. We are forced, therefore, to suppose that the prime mover conveys to the air (or water, or other such intermediary as is naturally capable both of moving and conveying motion) a power of conveying motion, but that this power is not exhausted when the intermediary ceases to be moved itself. Thus the intermediary will case to be moved itself as soon as the prime mover ceases to move it, but will still be able to move something else. Thus this something else will be put in motion after the prime mover's action has ceased and will itself continue the series. The end of it all will approach as the motive power conveyed to each successive secondary agent wanes, till at last there comes one which can only move its neighbor without being able to convey motive force to it. At this point the last active intermediary will cease to convey motion, the passive intermediary that has not active power will cease to be in motion, and the missile will come to a stand, at the same instant. Now this movement occurs in thing that are sometimes in motion and sometimes stationary, and it is not continuous, though it appears to be. For there is a succession of contiguous agents, since there is not one motor concerned but a series, one following upon another. And so there comes about both in air and water the kind of motion that some have called *antiperistasis*. But whereas the only possible solution of the problem it suggests is that which has just been explained, the theory of those who call it *antiperistasis* would involve the simultaneity of the action of every motor and the passion of every mobile in the series, and the simultaneity of their cessation. Whereas the fact is that the supposed continuity of the movement of the single mobile which sets us inquiring after the motor is only apparent; for in fact it is not impelled by one and the same motor throughout its course (Wicksteed and Cornford 1929).

But before proceeding to our conclusion it will be well to discuss a difficulty that arises in connexion with locomotion. If everything that is in motion with the exception of things that move themselves is moved by something else, how is it that some things, e.g. things thrown, continue to be in motion when their movent is no longer in contact with them? If we say that the movent in such cases moves something else at the same time, that the thrower e.g. also moves the air, and that this in being moved is also a movent, then it would be no more possible for this second thing than for the original thing to be in motion when the original movent is not in contact with it or moving it: all the things moved would have to be in motion simultaneously and also to have ceased simultaneously to be in motion when the original movent ceases to move them, even if, like the magnet, it makes that which it has moved capable of being a movent. Therefore, while we must accept this explanation to the extent of saying that the original movent gives the power of being a movent either to air or to water or to something else of the kind, naturally adapted for imparting and undergoing motion, we must say further that this thing does not cease simultaneously to impart motion and to undergo motion: it ceases to be in motion at the moment when its movent ceases to move it, but it still remains a movent, and so it causes something else consecutive with it to be in motion, and of this again the same may be said. The motion begins to cease when the motive force produced in one member of the consecutive series is at each stage less than that possessed by the preceding member, and it finally ceases when one member no longer causes the next member to be a movent but only causes it to be in motion. The motion of these last two—of the one as movent and of the other as moved—must cease simultaneously, and with this the whole motion ceases. Now the things in which this motion is produced are things that admit of being sometimes in motion and sometimes at rest, and the motion is not continuous but only appears so: for it is motion of things that are either successive or in contact, there being not one movent but a number of movents consecutive with one another: and so motion of this kind takes place in air and water. Some say that it is 'mutual replacement': but we must recognize that the difficulty raised cannot be solved otherwise than in the way we have described. So far as they are affected by 'mutual replacement', all the members of the series are moved and impart motion simultaneously, so that their motions also cease simultaneously: but our present problem concerns the appearance of continuous motion in a single thing, and therefore, since it cannot be moved throughout its motion by the same movent, the question is, what moves it? (Hardie and Gaye 1930).

But first it will be well to discuss a difficulty that arises in connexion with locomotion. If everything that is in motion with the exception of things that move themselves is moved by something, how is it that some things, e.g. things thrown, continue to be in motion when their mover is no longer in contact with them? If we say that the mover in such cases moves something else at the same time, e.g. the air, and that this in being moved is also a mover, then it will similarly be impossible for this to be in motion when the original mover is not in contact with it or moving it: all the things moved would have to be in motion simultaneously and also to have ceased simultaneously to be in motion when the original mover ceases to move them, even if, like the magnet, it makes that which it has moved capable of being a mover. Therefore, we must say that the original mover gives the power of being a mover either to air or to water or to something else of the kind, naturally adapted for imparting and undergoing motion; but this thing does not cease simultaneously to impart motion and to undergo motion: it ceases to be in motion at the moment when its mover ceases to move it, but it still remains a mover, and so it causes something else consecutive with it to be in motion, and of this again the same may be said. The motion ceases when the motive force produced in one member of the consecutive series is at each stage less, and it finally ceases when one member no longer causes the next member to be a mover but only causes it to be in motion. The motion of these last two—of the one as mover and of the other as moved—must cease simultaneously, and with this the whole motion ceases. Now the things in which this motion is produced are things that admit of being sometimes in motion and sometimes at rest, and the motion is not continuous but only appears so; for it is motion of things that are either successive or in contact, there being not one mover but a number consecutive with one another. That is why motion of this kind takes place in air and water. Some say that it is mutual replacement; but the difficulty raised cannot be solved otherwise than in the way we have described. Mutual replacement makes all the members of the series move and impart motion simultaneously, so that their motions also cease simultaneously; but there appears to be continuous motion in a single thing, and therefore, since it cannot be moved by the same mover, the question is, what moves it? (Barnes 1984).

But before going any further it would be a good idea to resolve a certain difficulty concerning movement. Given that, with the exception of self-movers, every moving object is moved by something, how is it that some things—things that are thrown, for instance—have continuity of movement when that which initiated the movement is no longer in contact with them? If the mover also causes something else to move—the air, for instance, which causes movement by being in motion itself—it remains equally impossible for the air to be in motion when the first cause of movement is no longer in contact with it or causing it to move. No, all the things that are moving must move at the same time as the first mover and must have stopped moving when the first mover stopped imparting motion, and this is so even if, like a loadstone, the first mover makes what it moved capable of causing movement itself. So what we have to say is that although the first cause of movement imparts the ability to cause movement to the air or the water (or whatever else it may be that is, by its nature, capable of causing movement and of being moved), nevertheless the air or water or whatever does not stop causing movement and being moved at the same time as the first mover stop; it my stop being moved as soon as the cause of movement stops imparting movement, but it retains its ability to cause movement. That is why it imparts movement to something else which is consecutive to it, and the same goes for this in turn. The process of stopping begins when each consecutive member of the series has less power to cause movement, and the motion finally comes to an end when the previous member of the series no longer makes the next one a cause of movement, but only makes it move. The movement of these last two members of the series, the mover and the moved, necessarily ends simultaneously, and so the whole movement comes to an end.

So this kind of movement occurs in things which are capable of sometimes being in motion and sometimes being at rest. Despite appearances, it is not continuous motion; for the objects are either successive or in contact, since no single mover is involved, but a number of movers, one after another. That is why this kind of movement, which some people call mutual replacement, occurs in air and water. But the way I have described is the only way to resolve the difficulty; mutual replacement fails to do this since it makes everything cause movement and be moved simultaneously, and so also stop simultaneously. As things are, though, the appearance is of a single thing which is moving continuously. So the question arises: by what is it being moved? And we find that it is not being moved by a single mover (Waterfield 1996).

First, however, it is well to raise a certain difficulty about things which are borne. For if everything moving is move by something, how are some things, like things thrown, which do not move themselves, moved continuously, not being touched by the mover? If the mover a the same time moves something else, like the air, which, being moved, moves [the other], it is similarly impossible that the air be moving, if the first thing does not touch or move [it], but all are moving at the same time and cease when the first mover will cease, even if the mover, like a magnet, makes what is moved albe to move. It is necessary, then, to say thins, that the first mover makes [the medium] able to move [another], either the air or the water or some other such thing which is naturally apt to move and to be moved, But id does not cease moving [he other] and being moved at the same time, but [it ceases] being moved when the one moving ceases moving [it], though it is still a mover. Whence also, it moves some other contiguous thing, and the account of this is the same. It ceases, however, when the power of moving [another] always comes to be less in the contiguous thing. At the end, when it no longer makes what is before it a mover, but only a thing moved, it ceases. It is, however, necessary that these things, the mover and the moved, and the whole motion, cease at the same time. This same motion, [i.e., projectile motion], then, comes to be in what can at times be moved and at times rest, and this motion is not continuous, but appears so. For it is [derived] from beings in succession or touching; for the mover is not one, but things contiguous to each other [are the movers]. Whence, such motion comes to be in air or water, which motion some men say is mutual replacement. But it is impossible to solve the difficulties raised otherwise than in the way said. Mutual replacement, however, makes all to be moved and to move [another] at the same time, so that it also makes them all cease [at the same time]. Now, however, the thing thrown appears as something one which is moved continuously. By what [is it moved], then? For not by itself (Coughlin 2005).[9]

On the Heavens, I.ii.268ᵇ14-24

All natural bodies and magnitudes we hold to be, as such, capable of locomotion; for nature, we say, is their principle of movement. But all movement that is in place, all locomotion, as we term it, is either straight or circular or a combination of these two, which are the only simple movements. And the reason of this is that these two, the straight and the circular line, are the only simple magnitudes. Now revolution about the centre is circular motion, while the upward and downward movements are in a straight line, 'upward' meaning motion away from the centre, and 'downward' motion towards it. All simple motion, then, must be motion either away from or towards or about the centre (Stock 1922).

...all natural bodies and magnitudes are capable of moving of themselves in space; for nature we have defined as the principle of motion in them. Now all motion in space (locomotion) is either straight or circular or a compound of the two, for these are the only simple motions, the reason being that the straight and circular lines are the only simple magnitudes. By 'circular motion' I mean motion around the centre, by 'straight,' motion up and down. 'Up' means away from the centre,' 'down' towards the centre. It follows that all simple locomotion is either away from the centre or towards the centre or around the centre (Guthrie 1939).

---

[9]οὐ γὰρ ὑπὸ τοῦ αὐτοῦ. Coughlin elected to translate "For not by itself," although the context seems to prefer "for not by the same thing."

On the Heavens, I.ii.268^{b}26-269^{a}6

Bodies are either simple or compounded of such; and by simple bodies I mean those which possess a principle of movement in their own nature, such as fire and earth with their kinds, and whatever is akin to them. Necessarily, then, movements also will be either simple or in some sort compound—simple in the case of the simple bodies, compound in that of the composite—and in the latter case the motion will be that of the simple body which prevails in the composition. Supposing, then, that there is such a thing as simple movement, and that circular movement is an instance of it, and that both movement of a simple body is simple and simple movement is of a simple body (for if it is movement of a compound it will be in virtue of a prevailing simple element), then there must necessarily be some simple body which revolves naturally and in virtue of its own nature with a circular movement (Stock 1922).

Of bodies some are simple, and some are compounds of the simple. By 'simple' I mean all bodies which contain a principle of natural motion, like fire and earth and their kinds, and the other bodies of the same order. Hence motions also must be similarly divisible, some simple and others compound in one way or another; simple bodies will have simple motions and composite bodies composite motions, though the movement may be according to the prevailing element in the compound.

If we take these premises, (a) that there is such a thing as simple motion, (b) that circular motion is simple, (c) that simple motion is the motion of a simple body[10] (for if a composite body moves with a simple motion, it is only by virtue of a simple body prevailing and imparting its direction to the whole), then it follows that there exists a simple body naturally so constituted as to move in circle in virtue of its own nature (Guthrie 1939).

On the Heavens, I.iii,269^{b}18-31

In consequence of what has been said, in part by way of assumption and in part by way of proof, it is clear that not every body either possesses lightness or heaviness. As a preliminary we must explain in what sense we are using the words 'heavy' and 'light', sufficiently, at least, for our present purpose: we can examine the terms more closely later, when we come to consider their essential nature. Let us then apply the term 'heavy' to that which naturally moves towards the centre, and 'light' to that which moves naturally away from the centre. The heaviest thing will be that which sinks to the bottom of all things that move downward, and the lightest that which rises to the surface of everything that moves upward. Now, necessarily, everything which moves either up or down possesses lightness or heaviness or both-but not both relatively to the same thing: for things are heavy and light relatively to one another; air, for instance, is light relatively to water, and water light relatively to earth. The body, then, which moves in a circle cannot possibly possess either heaviness or lightness. For neither naturally nor unnaturally can it move either towards or away from the centre (J.L. Stocks 1922).

After what has been said, whether laid down as hypothesis or demonstrated in the course of the argument, it becomes clear that not every body has either lightness or weight. However, we must first lay down what we mean by heavy and light, at present only so far as it is necessary for the purpose in hand, but later with more precision, when we come to

---

[10]Guthrie ignored the bi-directional structure of the Greek text to the effect that a simple body possesses simple motion and simple motion belongs to a simple body: καὶ τοῦ τε ἁπλοῦ σώματος ἁπλῆ ἡ κίνησις καὶ ἡ ἁπλῆ κίνησις ἁπλοῦ σώματος.

investigate the real nature of the two. Let 'the heavy' then be that whose nature it is to move towards the centre, 'the light' that whose nature it is to move away from the centre, 'heaviest' that which sinks below all other bodies whose motion is downwards, and 'lightest' that which rises to the top of the bodies whose motion is upwards. Thus every body which moves downwards or upwards must have either lightness or weight or both. (A body cannot of course be both heavy and light in relation to the same thing, but the elements are so in relation to each other, e.g. air is light in comparison with water, but water in comparison with earth.) Now the body whose motion is circular cannot have either weight or lightness, for neither naturally nor unnaturally can it ever move towards or away from the centre (Guthrie 1939).

On the Heavens, I.iii.269^{b}29-270^{a}12

The body, then, which moves in a circle cannot possibly possess either heaviness or lightness. For neither naturally nor unnaturally can it move either towards or away from the centre. Movement in a straight line certainly does not belong to it naturally, since one sort of movement is, as we saw, appropriate to each simple body, and so we should be compelled to identify it with one of the bodies which move in this way. Suppose, then, that the movement is unnatural. In that case, if it is the downward movement which is unnatural, the upward movement will be natural; and if it is the upward which is unnatural, the downward will be natural. For we decided that of contrary movements, if the one is unnatural to anything, the other will be natural to it. But since the natural movement of the whole and of its part of earth, for instance, as a whole and of a small clod-have one and the same direction, it results, in the first place, that this body can possess no lightness or heaviness at all (for that would mean that it could move by its own nature either from or towards the centre, which, as we know, is impossible); and, secondly, that it cannot possibly move in the way of locomotion by being forced violently aside in an upward or downward direction. For neither naturally nor unnaturally can it move with any other motion but its own, either itself or any part of it, since the reasoning which applies to the whole applies also to the part (J.L. Stocks 1922)

Now the body whose motion is circular cannot have either weight or lightness, for neither naturally nor unnaturally can it ever move towards or away from the centre. (a) Naturally it cannot have rectilinear motion, because it was laid down that each simple body has only one natural motion, and therefore it would itself be one of the bodies whose natural motion is rectilinear. (b) But suppose it moves in a straight line contrary to its nature, then if the motion is downwards, upward motion will be its natural one, and *vice versa*; for it was one of our hypotheses that of two contrary motion, if one is unnatural the other is natural. Taking into account then the fact that a whole and its part move naturally in the same direction (as do e.g. all earth together an a small clod), we have established (a) that it has neither lightness nor weight, since otherwise it would have been able to move naturally either towards the centre or away from the centre, (b) tht it cannot move locally by being violently forced either up or down, for it is impossible for it to move, either naturally or unnaturally, with any other motion but its own, either itself as a whole or any of its parts, seeing that the same argument applies to whole and part (Guthrie 1939).

On the Heavens, III.ii.301^{b}20-30

...since movement is always due either to nature or to constraint, movement which is natural, as downward movement is to a stone, will be merely accelerated by an external force, while an unnatural movement will be due to the force alone. *In either case the air is as it were instrumental to the force.* For air is both light and heavy, and thus qua light produces upward motion, being propelled and set in motion by the force, and qua heavy produces a downward motion. *In either case the force transmits the movement to the body by first, as it were, impregnating the air.* That is why a body moved by constraint continues to move when that which gave the impulse ceases to accompany it. Otherwise, i.e. if the air were not endowed with this function, constrained movement would be impossible. *And the natural movement of a body may be helped on in the same way* (J.L. Stocks 1922, my italics).

On the Heavens, III.ii.301^{a}24-301^{b}1

We go on to show that there are certain bodies whose necessary impetus is that of weight and lightness. Of necessity, we assert, they must move, and a moved thing which has no natural impetus cannot move either towards or away from the centre. Suppose a body A without weight, and a body B endowed with weight. Suppose the weightless body to move the distance CD, while B in the same time moves the distance C, which will be greater since the heavy thing must move further. Let the heavy body then be divided in the proportion CE:CD (for there is no reason why a part of B should not stand in this relation to the whole). Now if the whole moves the whole distance CE, the part must in the same time move the distance CD. A weightless body, therefore, and one which has weight will move the same distance, which is impossible. And the same argument would fit the case of lightness (J.L. Stocks 1922).

On the Heavens, III.ii.301^{b}5-18

For there will be a force which moves it, and the smaller and lighter a body is the further will *a given force* move it. Now let A, the weightless body, be moved the distance CE, and B, which has weight, be moved in the same time the distance CD. Dividing the heavy body in the proportion CE:CD, we subtract from the heavy body a part which will in the same time move the distance CE, since the whole moved CD: for *the relative speeds of the two bodies will be in inverse ratio to their respective sizes.* Thus the weightless body will move the same distance as the heavy in the same time. But this is impossible. Hence, since the motion of the weightless body will cover a greater distance than any that is suggested, it will continue infinitely. It is therefore obvious that every body must have a definite weight or lightness (J.L. Stocks 1922, my italics).

Physics VII.v.250^{a}12-19

If, then, A moves B a distance C in a time D, it does not follow that E, being half of A, will in the time D or in any fraction of it cause B to traverse a part of C the ratio between which and the whole of C is proportionate to that between A and E—in fact it might well be that it will cause no motion at all; for it does not follow that, if a given motive power causes a certain amount of motion, half that power will cause motion either of any particular amount or in any length of time: otherwise one man might move a ship, since both the motive power of the ship-haulers and the distance that they all cause the ship to traverse are divisible into as many parts as there are men (Barnes 1984, slight revision of Hardie and Gaye 1930).

   so if A moves B as far as C in time D, then E, which is half of A, would not move B, in time D or any part of D, through any part of C in the same ratio to the whole as E is to A; for it will perhaps not move it at all. For it is not the case, if the whole strength moved something so much, that the half will move it either any amount in any time whatever; for one person could move a ship if the strength of the ship-haulers or the distance they all moved it were divided by the number of them (Sachs 1995).

   So if A moves B the distance C in D, E (i.e. half A) will not move B in D or any part of D over a part of the distance C which bears the same ratio to the whole of C as that which obtains between A ad E. It may well be that E will not move B at all. After all, the fact that a given power as a whole has moved an object such-and-such a distance does not mean that half the power will move it any distance in any amount of time. If it did, one man could move a ship, since the power of the halers and the distance which they all moved the ship together are divisible by the number of haulers (Waterfield 1996).

   If, then, A moves B in D so much, C, the half of A, E, will not move B [through] something of C proportional to the whole C as A to E in the time D nor in something of D. For, if it chanced, it will not move anything at all. For it does not follow that if the whole strength moved [it] so much, the half will [it] so much nor in any time. For one man would move a ship, if indeed the strength of the ship haulers and the length which all moved [the ship] are cut into the number of [of movers] (Coughlin 2005).

Physics, VIII.viii. 263^{b}12-23

It is true that in continuous time the point is common to the past and future and is one and the same numerically, though not in function, being the end of the one and the beginning of the other; but as regards the subject of change it always belongs to the future and not to the past state of that subject. For suppose the time is represented by A and B, and the dividing 'now' by C, and call the thing that suffers change D, and suppose that D is white during the whole of A and not-white during the whole of B; then at the instant C it will be both white and not-white; for if it really is white during the whole of A, it must be true that it is white at any instant of A, and in B it is not-white, and C is in both A and B. So we must not allow that it is white at every point of A, but only at every point of A except the terminating instant C. This instant already belongs to B; and if D occupied the whole time A in the process of becoming not-white or of ceasing to be white, either process was complete at the instant C (Wicksteed and Cornford 1929).

   It is true that the point is common to both times, the earlier as well as the later, and that, while numerically one and the same, it is theoretically not so, being the finishing-point of the one and the starting-point of the other: but so far as the thing is concerned it belongs to the later stage of what happens to it. Let us suppose a time ABG and a thing D, D being white in the time A and not-white in the time B. Then D is at the moment G white and not-white: for if we were right in saying that it is white during the whole time A, it is true to call it white at any moment of A, and not-white in B, and G is in both A and B. We must not allow, therefore, that it is white in the whole of A, but must say that it is so in all of it except the last moment G. G belongs already to the later period, and if in the whole of A not-white was in process of becoming and white of perishing, at G the process is complete (Hardie and Gaye 1930).

It is true that the point is common to both times, the earlier as well as the later, and that, while numerically one and the same, it is not so in definition, being the end of the one and the beginning of the other; but so far as the thing is concerned it always belongs to the later affection. Let us suppose a time ACB and a thing D, D being white in the time A and not white in the time B. Then D is at C white and not white; for if we were right in saying that it is white during the whole time A, it is true to call it white at any moment of A, and not white in B, and C is in both A and B. We must not allow, therefore, that it is white in the whole of A, but must say that it is so in all of it except the last now C. C already belongs to the later period, and if in the whole of A not white was becoming and white perishing, at C it had become or perished (Barnes 1984).

So although the point is common to both earlier and later time, and although it is the same numerically single point, it is conceptually different from itself (because it is the end of one period of time and the beginning of the other) and always belongs to the later affection of the object involved. Take a stretch of time ACB and an object D, which is white in time A, but not white in time B. In C, then, it is both white and not white. For if the object was white for the whole of A, then it is true to say that it is white in any part of A, and if it is not white for the whole of B, it is true to say that it is not white in any part of B, and C is a part of both A and B. The solution is not to grant that it is white for the whole stretch of time, but to say that it is white for the whole stretch of time except the final now, namely C, which is already part of the later stretch of time. Whether during the whole of A it was coming to be white or ceasing to be white,[11] these processes are complete at C. (Waterfield 1996)

The point, then, is common to both (to both the before and the after), and is one and the same in number; in account, however, it is not the same. For it is the end of this, but the beginning of that. In the thing, however, it is always of the later passion. Let he time in which be ACB, the thing, D. This, in time A, is white, in B, not white. So in C it is white and not white. For in whatever part of A it is true to say it is white, if it was white for this whole time, and in B not white. C, however, is in both. One must not grant [that it is white] in the whole [AC], but [only in all] except the final now, C. This, however, is already the later time. Even if it was coming to be not white and white was being destroyed in the whole A, it had come to be or was destroyed in C. (Coughlin 2005)

Physics, VI.vi.237ᵃ3-11

Again, (2) if what enables us to say that movement *has* taken place in the whole time OT— or generally in any period you please—is that we have taken the terminal 'now' (for the

---

[11]Waterfield elected to read with Simplicius: καὶ εἰ ἐγίγνετο λευκὸν καὶ εἰ ἐφθείρετο λευκόν
Bekker has: καὶ εἰ ἐγίγνετο οὐ λευκόν, καὶ εἰ ἐφθείρετο λευκὸν ἐν τῷ Α.

terminal 'now' is what defines a period, a period being what lies between two 'nows'), then movement may equally be said to have taken place in the other periods (OS and ST).[12] But the half (of OT) has its terminal point (S) in our division; accordingly movement will have taken place in the half or, generally, in any part you please; for wherever we divide the period, it will be limited by a 'now' coinciding with that division. And since any period of time is divisible, and what lies between the two 'nows' is time, it follows that anything that changes at all has already completed an unlimited succession of changes (Wicksteed and Cornford 1929).

Again, if by taking the extreme moment of the time-for it is the moment that defines the time, and time is that which is intermediate between moments-we are enabled to say that motion has taken place in the whole time ChRh or in fact in any period of it, motion may likewise be said to have taken place in every other such period. But half the time finds an extreme in the point of division. Therefore motion will have taken place in half the time and in fact in any part of it: for as soon as any division is made there is always a time defined by moments. If, then, all time is divisible, and that which is intermediate between moments is time, everything that is changing must have completed an infinite number of changes (Hardie and Gaye 1930).

Secondly, if it is our grasp of the limiting now of AB that allows us to say that the object has moved in the time AB as a whole, or in general in any part of AB (because it is the limiting now that defines a period of time, and time is what comes between nows), by the same token we could also say that it has moved in the other parts of AB as well. But by dividing AB we have provided half of it with a limiting now, and so the object will have moved in half of AB and in general in any given part of AB. For to make a division is simultaneously to provide a tie defined by those nows. So if any and every time is divisible, and if time is what lies between nows, then every changing object has completed an infinite number of changes (Waterfield 1996).

Moreover, if we say a mobile has moved in the whole time XR, or, generally, in any time whatever, by taking the / extreme now of the time (for this is what determines, and what is between the nows is time), it may be said that it has moved in the other nows similarly. Of the half, however, the division is the extreme; whence, in the half too it will have moved and, generally, in any one of the parts. For the time is always determined by the nows, together with the cut. It / then, every time is divisible and what is between the nows is time, everything which is changing will have changed with respect to infinite things (Coughlin 2005).

Physics, V.iv.228^{b}26-30

But sometimes the variation is neither in the form of the track, nor in the continuity or discontinuity of the time occupied, nor in the maintaining or reversing of the direction, bu in a quality of the motion itself; for the variation may be in its quickness or slowness, since a motion unform in speed may be called uniform, and varying in speed varying. It follwos that velocity is not special to any one genus of change (Wicksteed and Cornford 1929).

Sometimes it {whatever makes a motion irregular} is found neither in the place nor in the time nor in the goal but in the manner of the motion: for in some cases the motion is differentiated by quickness and slowness: thus if its velocity is uniform a motion is regular, if not it is irregular. So quickness and slowness are not species of motion nor do they

---

[12]That is, XK and KR respectively, associating letters OST with letters XKR that mark the time period in the Greek source.

constitute specific differences of motion, because this distinction occurs in connexion with all the distinct species of motion (Hardie and Gaye 1930).

...it {motion} may also differ by being (b) not in the place or in the time or in that toward which motion proceeds, but in the manner in which motion proceeds, for sometimes it is specified by fastenss and slowness (for if the motion proceeds with the same speed, it is said to be regular [uniform], but if not, then irregular [nonuniform]). Consequently, neither are fastness and slowness species or differentiae of motion, seeing that they follow all motions which differ with respect to species;... (Apostle 1969).

Alternatively, non-uniformity may occur not where the change takes place, or in the time, nor in the end-point, but in the way the change happens. For instance, there may be inconstancy of speed; any change which happens at the same speed is uniform, whereas any change where the speed differs is non-uniform. That is why quickness and slowness are neither species nor differentiae of change: it is because they are found in all the various species of change (Waterfield 1996).

Sometimes the difference is not in the 'where' nor in the 'when' nor in the 'that to which,' but in the 'how.' For sometimes it is determined by fastness and slowness. For that of which the speed is the same is regular, that of which it is not, irregular. Whence, speed and slowness are not species or differences of motion, because / they follow upon all [motions], which are different according to species (Coughlin 2005).

On the Heavens, II.vi. 288^{b}29-30

Equally impossible is perpetual acceleration or perpetual retardation. For such movement would be infinite and indefinite, but every movement, in our view, proceeds from one point to another and is definite in character (J.L. Stock 1922).

Yet nor is it {movement} capable of forever accelerating or of decelerating forever, since the movement will be unlimited and indefinite, and we maintain that all movement is from something to something, i.e. determined (Leggatt 1995).

On the Heavens, I.v.272^{a}7-8

Again, if from a finite time a finite time be subtracted, what remains must be finite and have a beginning. And if the time of a journey has a beginning, there must be a beginning also of the movement, and consequently also of the distance traversed (J.L. Stocks 1922).

Again, if a finite time be subtracted from a finite time, the remainder must also be finite and have a beginning. But if the time of the journey has a beginning, so also must the movement, and hence the distance which is traversed (Guthrie 1939).

Physics VIII.viii.262^{a}12-17; [...] 262^{b}23-263^{a}3

But the dominant consideration in declaring that movement along a straight line cannot be continuous is that the reversing of a movement involves stopping it. This is true not only of motion on a straight line but of motion on a circular track; for it is not the same thing to move on a circular track and to go on moving round and round it, since it is possible either to go on continuously or, when you have come to the point of departure, to turn back again.

We may convince ourselves that reversal of a movement involves stopping it, not only by observation, but by reasoning, [...] For suppose a mobile H moves as far as D and then turns back and moves down again: then it has made the one point at which it turned function both as a beginning and an end, and therefore as two. It must therefore have stopped there; it cannot have arrived at it and have departed from it simultaneously, since that would involve being there and not being there at the same instant. ... we cannot say ... that the mobile is only at the point D 'at' a sectional point of time, but has never 'arrived at it' or 'departed from it,' for the 'end' which it comes to must be an end in actuality, not in potentiality only. So a 'point between' the extremities of a continuous line is only potentially a beginning and an end, but this one is actual; it is both the finishing-point as regarded from above, and so also the end-point of one motion and the beginning-point of the other.

The conclusion is that the motion of the mobile that turns back upon a straight line must stop. It is impossible, therefore, that there should be, on a straight line, continuous movement that is everlasting (Wicksteed and Cornford 1929).

But what shows most clearly that rectilinear motion cannot be continuous is the fact that turning back necessarily implies coming to a stand, not only when it is a straight line that is traversed, but also in the case of locomotion in a circle (which is not the same thing as rotatory locomotion: for, when a thing merely traverses a circle, it may either proceed on its course without a break or turn back again when it has reached the same point from which it started). We may assure ourselves of the necessity of this coming to a stand not only on the strength of observation, but also on theoretical grounds. [...] For suppose H in the course of its locomotion proceeds to D and then turns back and proceeds downwards again: then the extreme point D has served as finishing-point and as starting-point for it, one point thus serving as two: therefore H must have come to a stand there: it cannot have come to be at D and departed from D simultaneously, for in that case it would simultaneously be there and not be there at the same moment. And here ... we cannot argue that H is at D at a sectional point of time and has not come to be or ceased to be there. For here the goal that is reached is necessarily one that is actually, not potentially, existent. Now the point in the middle is potential: but this one is actual, and regarded from below it is a finishing-point, while regarded from above it is a starting-point, so that it stands in these same two respective relations to the two motions. Therefore that which turns back in traversing a rectilinear course must in so doing come to a stand. Consequently there cannot be a continuous rectilinear motion that is eternal (Hardie and Gaye 1930).

But the impossibility of continuous movement on a straight line is shown particularly clearly by the fact that anything which reverses its direction has to come to a standstill, not only if it is moving on a straight line, but also if it is moving on a circle. Moving *in* a circle is different from moving *on* a circle, because in the latter case the object may either do the same movement again or return to its starting-point and reverse direction. Rational argument, as well as the evidence of the senses, will convince us that it is necessary to stop when reversing direction. [...] For if an object G was in motion towards D and then reversed its direction and travelled back down again, it did treat the extremity D as both an end and a beginning and make this single point two. That is why it is bound to have been at a standstill; it did not simultaneously reach D and leave it, because in that case it would have been there and not have been there in the same now. But ... we cannot say that G was at D at a division of time and did not reach D or leave it. After all, in this case G must come to an actual final point, not merely a potential one. What the mid-points are potentially, D is actually; it is an end-point on the journey from the bottom and a starting point on the journey from the top, and by the same token it is a starting-point for the downward movement and an end-point for the upward movement. Anything that reverses its direction on a straight line must, then, come to a standstill. So it is impossible for there to be continuous, eternal movement in a straight line (Waterfield 1996).

It is most apparent, however, that it is impossible that motion on a straight line be continuous because it is necessary that a thing turning back come to a stand, not only if it be borne on a straight line, but even if it be borne on a circle. For to be borne in a circle and according to a circle are not the same. For at times what is moving goes on in a line; at times, however, coming to the same place whence it was urged, it will again turn back. The conviction that it is necessary to come to a stand is not only due to sense but also to argument. [...] For if G be borne to D, and again, having turned back, be borne downward, that to which, the end D, was used as an end and as a beginning, the one point as two. Whence, it is necessary that G stand. And it does not come to be in D and come to be away from D at the same time. For at the same time, in the same now, it would be an not be there. ... for one cannot say that G is at D in the cut, though it did not come to be [there] nor come to be away from [there]. For it is necessary to come to an end in act, not in potency. The points in the middle, then, are in potency, but this one is in act, and from below it is an end, from above a beginning. And so, in the same way, [it is an end and a beginning] of the motions. So it is necessary that what turns back in a straight line stand. So there cannot be continuous, eternal motion on a straight line (Coughlin 2005).

Physics, V.iv.228ᵃ25

... and there are things that have no limiting extremes at all, and others whose limiting extremes, though called by the same name of 'end,' are of differing nature; for how can the 'end' of a walking come into contact with the 'end' of a line and become identical with it? (Wicksteed and Cornford 1929).

Now some things have no extremities at all: and the extremities of others differ specifically although we give them the same name of 'end': how should e.g. the 'end' of a line and the 'end' of walking touch or come to be one? (Hardie and Gaye 1930).

Now, some things do not have limits, and although other things do, their limits are different in form from one another and share only a name. How could the end of a line and the end of a walk be in contact or become one? (Waterfield 1996).

Of some things, there are no extremities, of some, there are, but they are different in species and equivocal. For how could the extremity of a line and of walking touch or come to be one? (Coughlin 2005).

Physics, IV.ix.217ᵃ21-30

... our statement is based on the assumption that there is a single matter for contraries, hot and cold and the other natural contrarieties, and that what exists actually is produced from a potential existent, and that matter is not separable from the contraries but its being is different, and that a single matter may serve for colour and heat and cold.

The same matter also serves for both a large and a small body. This is evident; for when air is produced from water, the same matter has become something different, not by acquiring an addition to it, but has become actually what it was potentially, and, again, water is produced from air in the same way, the change being sometimes from smallness to greatness, and sometimes from greatness to smallness (Hardie and Gaye 1930).

But we say that, from among the underlying things, there is one material of contraries, of hot and cold and of the other natural opposites, and from what *is* potentially comes what is at work, (ἐκ δυνάμει ὄντος ἐνεργίᾳ ὂν γίνεται)[13] while the material is not separate, though the being of it is different, and is one in number, as it might happen to skin to be the material of both hot and cold.

But there is also the same material of both a large and a small body. For clearly, whenever air comes into being from water, the same material became it without receiving anything else in addition, but what it was potentially it became actively (ἐνεργίᾳ ἐγένετο), and in turn water comes from air in the same way, turning at one time into largeness out of smallness, at another into smallness out of largeness (Sachs 1995).

Our position, however, based on considerations we have already established, is that opposites (hot and cold, and the other naturally existing oppositions) have a single underlying matter; that something actual comes to be from a state of potential; that while the matter is not separable, it is different in definition; and that numerically it is the same matter for the hot and for the cold, and, if it so happens, for color too.

Moreover, the matter of a body when it is large is the same as its matter when it is small. This is obviously so: when water turns into air, the same matter becomes something else. Nothing is added to it; al that happens is that something which formerly existed potentially comes to exist actually (Waterfiled 1996).

... we say, from the things laid down, that there is one material of contraries, of hot and cold and of the other natural contrarieties; and that being in act comes to be from being in potency; and that material is not separable but to be material is different; and that material is one in number [even] if it should chance to be [the material] of color and of heat and of cold.

The material of a large body and of a small body is the same. This is clear: for when air comes to be from water, it is not by taking something additional that the material came to be something else, but what was in potency came to be in act; and, again, water from air similarly; sometimes [the material changes] from smallness to greatness, sometimes from greatness to smallness (Coughlin 2005).

Physics, IV.ix.217^b11-19

And note that the dense is heavy, and the tenuous light. For density and tenuity are accompanied by two characteristics each, inasmuch as heavy and hard things are held to be

---

[13]Sach's translation, which for the most part maintains both simplicity and closeness to the Greek text, features a few terminological preferences that introduce what some readers may consider unnecessarily awkward turns of phrase. Justified as his critique of using "substance" for "ousia" in the Medieval Christian translations of Aristotle may be, the word "substance" has lost practically all of its medieval Christian connotations to modern secular readers. In context, it renders Aristotle's Greek no worse than Sachs's "thinghood." (Sachs, pp. 7–8, 253–254). Perhaps most striking is Sachs's aversion to the terms "act," "actual," and "actuality" which almost ubiquitously stand for "energeia" and "entelecheia." (Sachs, pp. 21–22, 31, 244–245). The words "actual" and "actually" probably do not quite convey the sense of Aristotle's "energeia" as indicating an active, operating state of being, or his sense of "entelecheia" as being in the state of actively possessing the goal. Sachs prefers "being at work" for energeia, and "being at work staying the same" for entelecheia. Still, Sachs's approach produces a particularly sharp dissonance here, where he translates "energeia" first as "at work", but then uses in the following sentence the more natural sounding "became actively."

dense, and light and soft ones tenuous. But in the case of iron and lead, the superior heaviness does not go with superior hardness (Wicksteed and Cornford 1929).

The dense is heavy, and the rare is light. [Again, as the arc of a circle when contracted into a smaller space does not acquire a new part which is convex, but what was there has been contracted; and as any part of fire that one takes will be hot; so, too, it is all a question of contraction and expansion of the same matter.] There are two types in each case, both in the dense and in the rare; for both the heavy and the hard are thought to be dense, and contrariwise both the light and the soft are rare; and weight and hardness fail to coincide in the case of lead and iron (Hardie and Gaye 1930).

Anything dense is heavy, anything rare is light. Each of them—rarity and density—is associated with two qualities. Both the heavy and the hard are thought to be dense, and conversely both the light and the soft are thought to be rare. Heaviness and hardness do not coincide, however, in the case of lead and iron (Waterfield 1996).

The dense is heavy, the rare is light. (Moreover, just as a circumference of a circle contracted to a smaller does not take on the concave, but what was there is contracted, and everything which one might take of a flame is hot, so too every contraction and expansion is of the same material.) For two things are [found] in each case, i.e., in the cases of the dense and the rare: for the dense seems to be both heavy and hard, and the rare the opposites, both light and soft. However, the heavy and the hard do not go together in lead and iron (Coughlin 2005).

On the Heavens, II.vi288^{a}17-22

If the movement is uneven, clearly there will be acceleration, maximum speed, and retardation, since these appear in all irregular motions. The maximum may occur either at the starting-point or at the goal or between the two; and we expect natural motion to reach its maximum at the goal, unnatural motion at the starting-point, and missiles midway between the two (J.L. Stocks 1922).

If it moves irregularly, it is clear that there will be acceleration, top-speed, and deceleration of its locomotion; for every irregular locomotion has a deceleration, an acceleration, and a top-speed. The top-speed is either at the point-from-which a thing moves, at the point-to-which it moves, or at the mid-point of the trajectory, as perhaps for things moving naturally it is at the point-to-which they move, for those moving counter-naturally the point-from-which they move, and for projectiles at the mid-point (Legatt 1995).

On the Heavens, II.iv 287^{b}5-14

But the surface of water is seen to be spherical if we take as our starting-point the fact that water naturally tends to collect in a hollow place-'hollow' meaning 'nearer the centre'. Draw from the centre the lines AB, AC, and let their extremities be joined by the straight line BC. The line AD, drawn to the base of the triangle, will be shorter than either of the radii. Therefore the place in which it terminates will be a hollow place. The water then will collect there until equality is established, that is until the line AE is equal to the two radii. Thus water forces its way to the ends of the radii, and there only will it rest: but the line which connects the extremities of the radii is circular: thereforethe surface of the water BEC is spherical (J.L. Stocks 1922).

Yet that the surface of water is like this {namely, spherical} is clear, if one takes it as an assumption that water is such as always to flow into what is more hollow, and that what is nearer the centre is more hollow. Let, then, the lines AB and AC be drawn from the centre, and let them be joined by line BC. The line AD drawn to the base, then, is smaller than those from the cnre therefore the place is more hollow. Inconsequence, the water will flow about it until it is equalized. Now line AE is equal to those from the centre. Consequently the water must reach the lines from the centre, since it will rest at that moment. Now, the line in contact with those from the centre is the circumference; therefore the surface BEC of the water is spherical (Legatt 1995).

On the Heavens, I.viii. 276^{b}22-25

To postulate a difference of nature in the simple bodies according as they are more or less distant from their proper places is unreasonable. For what difference can it make whether we say that a thing is this distance away or that? One would have to suppose a difference proportionate to the distance and increasing with it, but the form is in fact the same (J.L. Stocks 1922).

It is wrong to suppose that the elements have a different nature if they are removed less or more from their proper places; for what difference does it make to say that they are removed by this distance or that? They will differ in proportion, more as the distance increases, but the form will remain the same (Guthrie 1939).

On the Heavens, I.viii. 277^{a}27-32

This conclusion that local movement is not continued to infinity is corroborated by the fact that earth moves more quickly the nearer it is to the centre, and fire the nearer it is to the upper place. But if movement were infinite speed would be infinite also; and if speed then weight and lightness. For as superior speed in downward movement implies superior weight, so infinite increase of weight necessitates infinite increase of speed (J.L. Stocks 1922).

Further evidence for the finite character of local motion is provided by the fact that earth moves more quickly, the nearer it is to the centre, and fire, the nearer it is to the upper limit. If the movement were to infinity, its speed would be infinite also, and if the speed were infinite, the weight or lightness of the object would be infinite also; for just a s a body which, by reason of its speed, occupied a lower position than another, would owe that speed to its weigh so an infinite increase in weight would mean an infinite increase on the speed (Guthrie 1939).

Bibliography

Primary Sources

Aristotle. 1984. Mechanics. Trans. E.S. Forster. In *The complete works of Aristotle, the revised Oxford translation*. Volume two. ed. J. Barnes, 1299–1318. Princeton: Princeton University Press.

Aristotle. 2007. *The Mechanical Problems In The Corpus of Aristotle*. Trans. Thomas Nelson Winter. http://digitalcommons.unl.edu/classicsfacpub/68.

Aristotle. 1922. *On the Heavens*. Trans. J.L. Stocks. Oxford: Clarendon Press.

Aristotle. 1939. *On the Heavens*. Trans. W.K.C. Guthrie. Cambridge, Mass.: Harvard University Press.

Aristotle. 1995. *On the Heavens I & II*. Trans. Stuart Leggatt. Warminster, England: Aris & Phillips, Ltd.

Aristotle. 1929. *Physics*. Trans. Francis M. Cornford and Philip H. Wicksteed. Cambridge, MA.: Harvard University Press.

Aristotle. 1930. *Aristotle's Physics*, Trans. R.P. Hardie and R.K. Gaye. Oxford: Clarendon Press.

Aristotle. 1969. *Aristotle's Physics*. Trans. Hippocrates G. Apostle. Bloomington and London: Indiana University Press.

Aristotle. 1993. *Aristotle Physics: Books III and IV*. Trans. Edward Hussey. New impression with corrections and additions. Oxford: Clarendon Press.

Aristotle. 1995. *Aristotle's Physics: A Guided Study*. Trans. Joe Sachs. New Burnswick: Rutgers University Press.

Aristotle. 1996. *Physics*. Trans. Robin Waterfield. Oxford: Oxford University Press.

Aristotle. 2005. *Aristotle: Physics, or Natural Hearing*. Coughlin, Glen. South Bend, Indiana: St. Augustine's Press.

Euclid. 1956. *The Thirteen Books of The Elements*. Translated with introduction and commentary by Sir Thomas L. Heath. 2nd Edition. New York: Dover Publications, Inc.

Euclid. 2008. *Euclid's Elements of Geometry*. Edited and provided with a new English translation by Richard Fitzpatrick http://farside.ph.utexas.edu/Books/Euclid/Elements.pdf.

Galileo Galilei. 1960. *On Motion and On Mechanics*. Trans. I.E. Drabkin and S. Drake. Madison, Wisconsin: University of Wisconsin Press.

Plato. 1937. *Plato's Cosmology: The Timaeus of Plato Translated with a Running Commentary*. Trans. F.M. Cornford. London: Routledge and Kegan Paul Ltd.

Secondary Sources

Bloom, Allan. 1991. *The Republic of Plato*, Translated with Notes and an Interpretive Essay, 2nd Edition. Basic Books, a division of HarperCollins Publishers.

Guthrie, W.K.C. 1975. *The Greek Philosophers: From Thales to Aristotle*. New York: Harper Tourchbooks.

© The Author(s) 2015
I. Yavetz, *Bodies and Media*, SpringerBriefs in History of Science and Technology, DOI 10.1007/978-3-319-21263-0

Guthrie, W.K.C. 1981. *Aristotle: An Encounter*. Cambridge: Cambridge University Press.

Hussey, Edward. 1991. Aristotle's Mathematical Physics: A Reconstruction. In *Aristotle's physics: a collection of essays*, ed. Lindsay Judson, 213–242. Oxford: Clarendon Press.

McLaughlin, Peter. 2013. *The Question of the Authenticity of the Mechanical Problems*. http://www.philosophie.uni-hd.de/md/philsem/personal/mclaughlin_authenticity_2013_2.pdf.

Thorp, J. 1982. The Luminousness of the Quintessence. *Phoenix* 36: 104–123.

Wolff, Michael. 1987. Philoponus and the Rise of Preclassical Dynamics. In *Philoponus and the Rejection of Aristotelian Science*, ed. R. Sorabji, 84–120. London: Gerald Duckworth & Co., Ltd.

Valleriani, Matteo. 2010. *Galileo Engineer*. Dordrecht: Springer.

Printed in the United States
By Bookmasters

Printed in the United States
By Bookmasters